FORESTS AND GRASSLANDS

THE LIVING EARTH

FORESTS AND GRASSLANDS

EDITED BY JOHN P. RAFFERTY, ASSOCIATE EDITOR, EARTH AND LIFE SCIENCES

IN ASSOCIATION WITH

Published in 2011 by Britannica Educational Publishing
(a trademark of Encyclopædia Britannica, Inc.)
in association with Rosen Educational Services, LLC
29 East 21st Street, New York, NY 10010.

Copyright © 2011 Encyclopædia Britannica, Inc. Britannica, Encyclopædia Britannica, and the Thistle logo are registered trademarks of Encyclopædia Britannica, Inc. All rights reserved.

Rosen Educational Services materials copyright © 2011 Rosen Educational Services, LLC. All rights reserved.

Distributed exclusively by Rosen Educational Services.
For a listing of additional Britannica Educational Publishing titles, call toll free (800) 237-9932.

First Edition

Britannica Educational Publishing
Michael I. Levy: Executive Editor
J.E. Luebering: Senior Manager
Marilyn L. Barton: Senior Coordinator, Production Control
Steven Bosco: Director, Editorial Technologies
Lisa S. Braucher: Senior Producer and Data Editor
Yvette Charboneau: Senior Copy Editor
Kathy Nakamura: Manager, Media Acquisition
John P. Rafferty: Associate Editor, Earth Sciences

Rosen Educational Services
Alexandra Hanson-Harding: Editor
Nelson Sá: Art Director
Cindy Reiman: Photography Manager
Matthew Cauli: Designer, Cover Design
Introduction by Jeanne Nagle

Library of Congress Cataloging-in-Publication Data

Forests and grasslands / edited by John P. Rafferty.
   p. cm. (The living earth)
"In association with Britannica Educational Publishing, Rosen Educational Services."
Includes bibliographical references and index.
ISBN 978-1-61530-313-7 (library binding)
1. Forests and forestry. 2. Grasslands. I. Rafferty, John P.
SD373.F62 2011
578.73—dc22

2010019061

Manufactured in the United States of America

On the cover: Great Spotted Woodpecker (*Dendrocopos Major*), Baarn, The Netherlands. *Gertjan Hooljer*

On page x: Mixed evergreen and hardwood forest on the slopes of the Adirondack Mountains near Keene Valley, New York. *Jerome Wyckoff*

On pages x, 1, 133, 178, 226, 227, 229, 235: Boreal forest, U.S., dominated by spruce trees (*Picea*). *Erwin & Peggy Bauer/Bruce Coleman Ltd.*

Contents

Introduction   x

Chapter 1: Forests   1
  Rainforests   4
  The Tropical Rainforest   7
    The Origin of the Tropical Rainforest   8
    Tropical Rainforest Environments   14
    The Biota of Tropical Rainforests   18
    The Monsoon Forest   20
    Life in a Bromeliad Pool   22
    The Cloud Forest   25
    Population and Community Development and Structure of Tropical Rainforests   25
    "Flying" Trees: Aerial Seed Dispersal in the Panamanian Rainforest   27
    Rainforest Regeneration in Panama   34
    Apartments of the Rainforest: Communities in Tree Hollows   35
    Hitching a Ride: Seed Dispersal by Animals in the Panamanian Rainforest   37
    The Biological Productivity of Tropical Rainforests   41
    Eating the Rainforest: Herbivory and How Plants Defend Themselves   42
    The Status of the World's Tropical Forests   44
  The Temperate Forest   53
    The Origin of the Temperate Forest   54

Temperate Forest Environments	56
The Deciduous Forest	60
The Thorn Forest	68
Population and Community Development and Structure of Temperate Forests	69
The Biological Productivity of Temperate Forests	74
The Boreal Forest	75
The Origin of the Boreal Forest	75
The Distribution of the Boreal Forest	77
The Climate of the Boreal Forest	79
The Coniferous Forest	81
Boreal Forest Soils	83
The Biota and Its Adaptations	86
The Community Structure of Boreal Forests	98
The Biological Productivity of Boreal Forests	101
Deforestation	102
Notable Forests of the World	106
The Amazon and the Ituri	106
Other Forests	118

Chapter 2: Forestry	**133**
The History of Forestry	133
The Ancient World	133
Medieval Europe	134
Modern Developments	135
The Development of U.S. Policies	137
The Classification and Distribution of Forests	140
Gymnosperms	141
Angiosperms	141

The Occurrence and Distribution of Forests	142
The Purposes and Techniques of Forest Management	147
Multiple-Use Concept	147
Sustained Yield	148
Forest Products	149
Silviculture	150
Range and Forage	159
Recreation and Wildlife	160
Watershed Management and Erosion Control	164
Fire Prevention and Control	166
Forest Fires	167
Insect and Disease Control	173
Agroforestry	176
Urban Forestry	177
Chapter 3: Grasslands	**178**
The Origin of Grasslands	178
Grassland Climates and Soils	182
Grassland Biota	183
Grassland Population and Community Development and Structure	186
The Biological Productivity of Grasslands	188
Grasses	189
Prairies	190
Savannas	192
The Origin of Savannas	193
Savanna Climates and Soils	195
The Biota of Savannas	197
The Population and Community Development and Structure of Savannas	201
The Biological Productivity of	

Savannas	204
Chaparral	206
Tundras	206
The Arctic Tundra	207
The Alpine Tundra	210
Notable Grasslands of the World	211
Buffalo Gap National Grassland	211
The Llanos	212
The Nyika Plateau	214
Oglala National Grassland	214
Serengeti National Park	215
Veld	217
Conclusion	223
Glossary	226
Bibliography	228
Index	234

INTRODUCTION

Introduction

When attention to the smallest detail detracts from a person's comprehension of the larger situation at hand—"the big picture"—it is said that he or she "can't see the forest for the trees." In the strictest sense, this idiom misses the mark. Ecologists and other biogeographic scientists, however, could easily argue that the only way to truly see a forest is to pay attention to the trees themselves and the activity occurring between each one.

Ecosystems are geographic areas that are home to similar biota (plants and animals of an area) that have adapted to a particular region's environmental conditions. Forest ecosystems are dominated by trees, and grasslands are even easier to identify, as they are dominated by species of wild grasses.

This book takes readers deep inside a variety of forests and out across the sweeping vistas created by various grasslands. In the process, they will encounter the elements inherent to each particular ecosystem, study the climatic conditions that surround these areas, and absorb the importance of each of these biomes to the overall balance of life on Earth. General observations will be supported through an examination of specific, notable forests and grasslands around the globe.

Forests may be divided into a number of different categories, each defined by the type of trees they contain. Soil quality—chiefly as it pertains to the abundance of nutrients—the availability of sunlight, the amount of precipitation the soil receives, and the range of temperature the soil experiences determine which tree species grow in a particular area. Warm weather, high humidity, and copious amounts of rainfall favour the trees of the tropical rainforest, a highly diverse range of plants that includes palms, cycads, and the valuable hardwood teak, among others. In fact, the world's tropical rainforests, which are located near the Equator, contain the most biological diversity of any ecosystem on the planet. High levels of rain and humidity tend to leach many nutrients from the soil in tropical rainforests, and the remaining minerals tend to be bound up in the plants themselves. What nutrients remain in the soil are concentrated mainly in the top layer. Consequently, trees with shallow root systems do well in this type of ecosystem.

Trees of the rainforest are typically tall, reaching heights of 30 to 50 metres (98 to 164 feet). They generally grow close together, and their broad, rubbery leaves overlap high above the forest floor to form the canopy. A large number of species have adapted themselves to live among the lofty branches of the rainforest canopy: a diverse collection of insects, colourful parrots and birds of paradise, and tree-dwelling herbivores such as monkeys and flying squirrels. Smaller trees also populate the rainforest in an area called the understory, growing in whatever light gets filtered through the leafy roof of the canopy. The forest floor is covered with decomposing plant and animal matter and debris. Each of these forest layers (the canopy, the understory, and the forest floor) is home to diverse groups of animals that specialize on the opportunities provided by these unique environments.

In contrast boreal, or taiga, forests occur in cold regions. The word *boreal* translates to "northern," and climates in the north are notoriously harsh. The word taiga (Russian: "little sticks") originally referred to Russia's northern forests, but it is also used to describe boreal forests. The boreal forest is dominated by conifers, which have multiple small needles that store nutrients for longer periods than do broad leaves. Alternatively, broad-leaved trees depend on longer periods of warmth to maintain optimal rates of photosynthesis, a process that converts sunlight to chemical energy. As a result, broad-leaved trees are not fit to survive the long periods of darkness and cold characteristic of subarctic regions. The list of conifers includes firs, pines, spruces, and other species associated with the term *evergreen*. Aspen, birch, and larch, which have smallish, thin leaves with sharp edges or teeth, may also occur in some boreal forests.

As one might imagine, winters are long and quite cold in the boreal forest. The growing season in this region is confined to a few months in the summer, with temperatures that, on average, don't reach much above 20 °C (68 °F). Consequently, permafrost is found over wide areas of the soil contained within these forests. Defined roughly as consistently or permanently frozen ground that lies below the ground's surface crust, permafrost inhibits the root systems of vegetation in boreal forests, thus allowing trees, shrubs, and plants that can thrive with shallow roots to dominate. Tundra regions, characterized by low, shrubby plants occur poleward of boreal forests.

When considering the range of forest biodiversity, the boreal forest is relatively limited compared to the richly diverse rainforests of the tropics. The species of trees are few, but boreal forests contain a wide range of mosses and lichens that cover the forest floor. Other plants that grow in regions close to and within the Arctic Circle include quick-growing fireweed, orchids such as coral root and lady's slipper, that grow in association with fungi; and berries, such as cloudberry and lingonberry plants.

Compared to the teeming faunas of the tropical rainforests, relatively few species of mammals, birds, and insects live in taiga habitats because of the challenging climatic conditions. Those that do occur have found ways to adapt to this forbidding environment. Many warm-blooded mammals, such as the snowshoe hare and the lynx, generally have thick, heavy fur that keeps them warm. The coats of some of these animals also act as camouflage, changing from an earthy brown in the summer to a snowy white in the winter. To help them maneuver in terrain rife with ice and snow drifts, many taiga creatures possess disproportionately large feet that function like snowshoes, preventing them from sinking into the heavy snow cover.

Migration is another adaptation useful in the boreal forest. Hundreds of bird species appear in the taiga during the summer months. However, most of those do not live there year-round. Many birds that live in southern climates during the winter travel north to the taiga because the air is thick with insects to eat during the warm growing season. Reindeer also migrate, moving south to find food during the winter and returning to the forest as temperatures warm.

Between the hot, humid rainforests near the Equator and the boreal forests near the North Pole are temperate forests. Life in a temperate forest is marked by a moderate climate. These wooded areas, found primarily in North America, eastern Asia, and western Europe, are filled with stands of deciduous trees such as oaks, hickories, and maples, which sport broad-leaved foliage. Australian and Mediterranean temperate forests commonly have sclerophyllous trees (which are characterized by thick, hard leaves), such as Eucalyptus trees.

Unlike conifers, which are able to store nutrients year-round and thus remain "evergreen," deciduous trees are subject to the temperature and moisture fluctuations of the seasons. During winter months, temperature declines. Reduced sunlight and available water translate into lower

Introduction

rates of photosynthesis. The leaves of deciduous trees turn different colours and fall off as the production of chlorophyll, a chemical pigment involved in photosynthesis, wanes.

During the warm growing season in temperate regions, a number of plants flower and leaf. Also growing in the shade beneath the canopy of leaves are saplings—the next generation of trees to populate the temperate forest. Many plants, saplings, and the berries or flowers they produce serve as food for the broad array of animals and birds that live in this region. The fauna of any particular forest generally reflects the species that are indigenous to the surrounding, nonforested region. Therefore, species composition varies according to the location of the forest. Because of the abundance of trees, however, all temperate woodlands are home to several kinds of birds and arboreal, or tree-dwelling, mammals. The latter includes squirrels in North America and certain species of monkeys in Asia.

Nearly one-third of Earth's surface is occupied by forests of one type or another. Common sense, then, would dictate that forested regions are an important component of the biosphere—the part of the Earth system that supports life—and therefore are worthy of humankind's attention and respect. Managing, preserving, and studying the world's woodlands is the purview of those undertaking the science of forestry. An ancient activity practiced, to some extent, since at least the height of the Roman Empire, forestry has evolved from simply a way to ensure a steady supply of wood to an ecological and cultural imperative in many parts of the world. Forests provide timber, recreational opportunities, a defense against soil erosion, habitat for wildlife, and other environmental services. Forestry is designed to protect and perpetuate valuable woodland resources.

— Forests and Grasslands —

Covering vast portions of Earth's land area, grassland ecosystems are an important natural resource. Like forests, grasslands serve as habitat for wildlife, but they are also used as grazing land for wild animals and farm animals alike. There are many differences between grasslands and the forests they are adjacent to. Physically, forests are complex ecosystems with multiple levels, whereas grasslands appear to be more simplified environments. The effects of wind and light penetration are often greater and overall humidity and soil moisture are lower in grasslands than in forests. Although some animals and other organisms utilize both of these ecosystems, many prefer one over the other.

Considered a type of intermediate ecosystem between the lush growth of forests and arid, veritably barren deserts, there are many types of grasslands on Earth's surface. They are often found in areas characterized by relatively drier conditions compared to nearby forests. Temperate grasslands are found in North America and Eurasia. Tropical grasslands exist in East Africa, Australia, and South America, among other tropical and subtropical locales.

In North America, there are wide expanses of grassland commonly referred to as prairies. The grasses and flowering plants that dominate these areas are perennial, meaning they come back year after year during the growing season. Rich soil and humid conditions give rise to tallgrass prairies, where the cordgrass grows high and seems to undulate like waves when touched by a breath of wind. The most prevalent type of prairie in North America is the midgrass prairie, an ecosystem that includes porcupine, buffalo, and grama grasses, along with wheatgrass. Variations of many of these same types of grasses take root on a shortgrass prairie, only drier soil stunts their growth.

Introduction

The type of vegetation that dominates tropical grasslands depends largely on how much precipitation that area receives. Semi-arid regions such as the Sahel grassland in Africa, which lies southwest of the Sahara desert, are covered by species such as *Aristida*, *Cenchrus*, and *Schoenefeldia*. Grassy swaths in wetter East Africa are dominated by the *Pennisetum* and *Hyparrhenia* grass families, and Australia's hot, arid climate supports large areas of spinifex *Plectrachne* and *Triodia* grasses.

With regard to fauna, grasslands generally attract an assortment of herbivores, as well as the occasional carnivore that uses the ecosystem as a hunting ground. The North American prairie includes coyote, jackrabbits, bison, hawks, grasshoppers, and rodents called prairie dogs. Residents of the Pampas in South America include small mammals such as foxes and skunks, herds of llama-like guanaco, and several bird species. Kangaroos inhabit Australian grasslands, as well as camels, wild horses, and other creatures that were once domesticated or have been introduced from other lands.

The tropical savanna occurs in hot, dry climates such as those found in parts of Africa, South America, India, and Madagascar. Various grasses dominate the savanna landscapes, but these regions also support a smattering of trees, shrubs, and woody plants. The rough-leaved deciduous raspa tree and the flowering Nance and cork trees dot South American savannas, growing amid rice grass (*Leersia*) and exceedingly tall *Paspalum*, which are important plants in these areas. Palms and baobabs, the latter with their distinctively twisted trunks, are interspersed among the grasses of the East African savanna. West Africa has expanses of high elephant grass, broken by the occasional canopy created by bushwillows and Anogeissus trees.

Many creatures great and small, wild and "kept" as agricultural livestock, roam grassland environments. These include large grazers indigenous to Africa—elephants, antelopes, rhinoceroses, and zebras—as well as invertebrates such as termites, which consume and compost dead plant material, thus enriching the surrounding soil. Grasslands also contain domestic animals such as cattle raised in the North American prairies, Eurasian steppe, and Argentina's Pampas.

Ecosystems such as forests and grasslands can be viewed as little worlds unto themselves, each with their own climates, inhabitants, interactions, and life cycles. However, each of these worlds interacts with one another to contribute to Earth's biosphere. Careful attention to both ecosystem's defining characteristics—trees in forests, grasses in grasslands—and the wider scope of an area is the best way to appreciate the worth and complex beauty of these systems.

CHAPTER 1

Forests

Forests are complex ecological systems in which trees are the dominant life-form. Together, forests of all types cover nearly 30 percent of Earth's land surface. Tree-dominated forests can occur wherever the temperatures rise above 10°C (50°F) in the warmest months and the annual precipitation is more than 200 mm (8 inches). They can develop under a variety of conditions within these climatic limits, and the kind of soil, plant, and animal life differs according to the extremes of environmental influences. In cool, high-latitude subpolar regions, forests are dominated by hardy conifers like pines, spruces, and larches. These taiga (boreal) forests have prolonged winters and between 250 and 500 mm (10 and 20 inches) of rainfall annually.

Reindeer (Rangifer tarandus), such as this grazer looking for grass near Torinen, Sweden, are well adapted for life in taiga forests in northern Europe and Asia. Olivier Morin/AFP/Getty Images

In more temperate high-latitude climates, mixed forests of both conifers and broad-leaved deciduous trees predominate. Broad-leaved deciduous forests develop in middle-latitude climates, where there is an average temperature above 10 °C (50 °F) for at least six months every year and annual precipitation is above 400 mm (16 inches). A growing period of 100 to 200 days allows deciduous forests to be dominated by oaks, elms, birches, maples, beeches, and aspens. In the humid climates of the equatorial belt, tropical rainforests develop. There heavy rainfall supports evergreens that have broad leaves instead of needle leaves, as in cooler forests. In the lower latitudes of the Southern Hemisphere, the temperate deciduous forest reappears.

Forest types are distinguished from each other according to species composition (which develops in part according to the age of the forest), the density of tree

Mixed evergreen and hardwood forest on the slopes of the Adirondack Mountains near Keene Valley, New York. Jerome Wyckoff

cover, types of soils found there, and the geologic history of the forest region.

Soil conditions are distinguished according to depth, fertility, and the presence of perennial roots. Soil depth is important because it determines the extent to which roots can penetrate into the earth and, therefore, the amount of water and nutrients available to the trees. The soil of taiga forests is sandy and quickly drained. Deciduous forests have brown soil, richer than sand in nutrients, and less porous. Rainforests and savanna woodlands have a soil layer rich in iron or aluminum, which give the soils either a reddish or yellowish cast. The amount of water available to the soil, and therefore available for tree growth, depends on the amount of annual rainfall. Water may be lost by evaporation from the surface or by leaf transpiration. Evaporation and transpiration also control the temperature of the air in forests, which is always slightly warmer in cold months and cooler in warm months than the air in surrounding regions.

The density of tree cover influences the amount of both sunlight and rainfall reaching every forest layer. A full-canopied forest absorbs between 60 and 90 percent of available light, most of which is absorbed by the leaves for photosynthesis. The movement of rainfall into the forest is considerably influenced by leaf cover, which tends to slow the velocity of falling water, which penetrates down to the ground level by running down tree trunks or dripping from leaves. Water not absorbed by the tree roots for nutrition runs along root channels, so water erosion is therefore not a major factor in shaping forest topography.

Forests are among the most complex ecosystems in the world, and they exhibit extensive vertical stratification. Conifer forests have the simplest structure: a tree layer rising to about 30 metres (98 feet), a shrub layer that is spotty or even absent, and a ground layer covered with lichens, mosses, and liverworts. Deciduous forests are more complex; the tree canopy is divided into an upper and lower story, while rainforest canopies are divided into at least three strata. The forest floor in both of these forests consists of a layer of organic matter overlying mineral soil. The humus layer of tropical soils is affected by the high levels of heat and humidity, which quickly decompose whatever organic matter exists. Fungi on the soil surface play an important role in the availability and distribution of nutrients, particularly in the northern coniferous forests. Some species of fungi live in partnership with the tree roots, while others are parasitically destructive.

Animals that live in forests have highly developed hearing, and many are adapted for vertical movement through the environment. Because food other than ground plants is scarce, many ground-dwelling animals use forests only for shelter. In temperate forests, birds distribute plant seeds and insects aid in pollination, along with the wind. In tropical forests, fruit bats and birds effect pollination. The forest is nature's most efficient ecosystem, with a high rate of photosynthesis affecting both plant and animal systems in a series of complex organic relationships.

RAINFORESTS

A rainforest is a luxuriant forest, generally composed of tall, broad-leaved trees and usually found in wet tropical uplands and lowlands around the Equator.

Rainforests usually occur in regions where there is a high annual rainfall of generally more than 1,800 mm (70 inches) and a hot and steamy climate. The trees found in these regions are evergreen. Rainforests may also be found in areas of the tropics in which a dry season occurs, such as the "dry rainforests" of northeastern Australia. In these regions annual rainfall is between 800 and 1,800 mm (31 and 70 inches) and as many as 75 percent of the trees are deciduous.

Tropical rainforests are found primarily in South and Central America, West and Central Africa, Indonesia, parts of Southeast Asia, and tropical Australia. The climate in these regions is one of relatively high humidity with no marked seasonal variation. Temperatures remain high, usually about 30 °C (86 °F) during the day and 20 °C (68 °F) at night. Where altitude increases along the borders of equatorial rainforests, the vegetation is replaced by montane forests, as in the highlands of New Guinea, the Gotel Mountains of Cameroon, and in the Ruwenzori mass of Central Africa. Tropical deciduous forests are located mainly in eastern Brazil, southeastern Africa, northern Australia, and parts of Southeast Asia.

Other kinds of rainforests include the monsoon forests, most like the popular image of jungles, with a marked dry season and a vegetation dominated by deciduous trees such as teak, thickets of bamboo, and a dense undergrowth. Mangrove forests occur along estuaries and deltas on tropical coasts. Temperate rainforests filled with evergreen and laurel trees are lower and less dense than other kinds of rainforests because the climate is more equable, with a moderate temperature range and well-distributed annual rainfall.

The topography of rainforests varies considerably, from flat lowland plains marked by small rock hills to highland valleys crisscrossed by streams. Volcanoes that produce rich soils are fairly common in the humid tropical forests.

Soil conditions vary with location and climate, although most rainforest soils tend to be permanently moist and soggy. The presence of iron gives the soils a reddish or yellowish colour and develops them into two types of soils: extremely porous tropical red loams, which can be easily tilled; and lateritic soils, which occur in well-marked layers that are rich in different minerals. Chemical weathering of rock and soil in the equatorial forests is intense, and in rainforests weathering produces soil mantles up to 100 metres (330 feet) deep. Although these soils are rich in aluminum, iron oxides, hydroxides, and kaolinite, other minerals are washed out of the soil by leaching and erosion. The soils are not very fertile, either, because the hot, humid weather causes organic matter to decompose rapidly and to be quickly absorbed by tree roots and fungi.

Rainforests exhibit a highly vertical stratification in plant and animal development. The highest plant layer, or tree canopy, extends to heights between 30 and 50 metres (98 to 164 feet). Most of the trees are dicotyledons, with thick leathery leaves and shallow root systems. The nutritive, food-gathering roots are usually no more than a few centimetres deep. Rain falling on the forests drips down from the leaves and trickles down tree trunks to the ground, although a great deal of water is lost to leaf transpiration.

Most of the herbaceous food for animals is found among the leaves and branches of the canopy, where a variety of animals have developed swinging, climbing, gliding, and leaping movements to seek food and escape predators. Monkeys, flying squirrels, and sharp-clawed woodpeckers are some of the animals that inhabit the treetops. They rarely need to come down to ground level.

The next lowest layer of the rainforest is filled with small trees, lianas, and epiphytes, such as orchids, bromeliads, and ferns. Some of these are parasitic, strangling their host's trunks; others use the trees simply for support.

Above the ground surface the space is occupied by tree branches, twigs, and foliage. Many species of animals run, flutter, hop, and climb in the undergrowth. Most of these animals live on insects and fruit, although a few are carnivorous. They tend to communicate more by sound than by sight in this dense forest strata.

Contrary to popular belief, the rainforest floor is not impassable. The ground surface is bare, except for a thin layer of humus and fallen leaves. The animals inhabiting this strata, such as rhinoceroses, chimpanzees, gorillas, elephants, deer, leopards, and bears, are adapted to walking and climbing short distances. Below the soil surface, burrowing animals, such as armadillos and caecilians, are found, as are microorganisms that help decompose and free much of the organic litter accumulated by other plants and animals from all strata.

The climate of the ground layer is unusually stable. The upper stories of tree canopies and the lower branches filter sunlight and heat radiation, as well as reduce wind speeds, so that the temperatures remain fairly even throughout the day and night.

Virtually every group of animals except fishes is represented in the rainforest ecosystem. Many invertebrates are very large, such as giant snails and butterflies. The breeding seasons for most animals tend to be coordinated with the availability of food, which, although generally abundant, does vary seasonally from region to region. Climatic variations, however, are slight and thus affect animal behaviour very little. Those animals that do not have highly developed modes of quick locomotion are concealed from predators by camouflage or become nocturnal feeders.

THE TROPICAL RAINFOREST

Although similar to temperate rainforests, tropical rainforests are found in wet tropical uplands and lowlands around the Equator.

Rainforests are vegetation types dominated by broad-leaved trees that form a dense upper canopy (layer of foliage) and contain a diverse array of vegetation. Contrary to common thinking, not all rainforests occur in places with high, constant rainfall; for example, in the so-called "dry rainforests" of northeastern Australia the climate is punctuated by a dry season, which reduces the annual precipitation. Nor are all forests in areas that receive large amounts of rainfall true rainforests; the conifer-dominated forests in the extremely wet coastal areas of the American Pacific Northwest are temperate evergreen forest ecosystems. Therefore, to avoid conveying misleading climatic information, the term *rainforest* is now preferred over *rain forest*.

This section covers only the richest of rainforests — the tropical rainforests of the ever-wet tropics.

THE ORIGIN OF THE TROPICAL RAINFOREST

Tropical rainforests represent the oldest major vegetation type still present on the terrestrial Earth. Like all

Rainforest vegetation along the northern coast of Ecuador. © Victor Englebert

vegetation, however, that of the rainforest continues to evolve and change, so that modern tropical rainforests are not identical with rainforests of the geologic past.

Tropical rainforests grow mainly in three regions: the Malesian botanical subkingdom, which extends from Myanmar (Burma) to Fiji and includes the whole of Thailand, Malaysia, Indonesia, the Philippines, Papua New Guinea, the Solomon Islands, and Vanuatu and parts of Indochina and tropical Australia; tropical South and Central America, especially the Amazon basin; and West and Central Africa. Smaller areas of tropical rainforest occur elsewhere in the tropics wherever climate is suitable. The principal areas of tropical deciduous forest are in India, the Myanmar–Vietnam–southern coastal China region, and eastern Brazil, with smaller areas in South and Central America north of the Equator, the West Indies, southeastern Africa, and northern Australia.

The flowering plants (angiosperms) first evolved and diversified during the Cretaceous Period about 100 million years ago, during which time global climatic conditions were warmer and wetter than those of the present. The vegetation types that evolved were the first tropical rainforests, which blanketed most of the Earth's land surfaces at that time. Only later—during the middle of the Paleogene Period, about 40 million years ago—did cooler, drier climates develop, leading to the development across large areas of other vegetation types.

It is no surprise, therefore, to find the greatest diversity of flowering plants today in the tropical rainforests where they first evolved. Of particular interest is the fact that the majority of flowering plants displaying the most primitive characteristics are found in rainforests (especially tropical rainforests) in parts of the Southern Hemisphere, particularly South America, northern Australia and adjacent regions of Southeast Asia, and some

Vegetation profile of a tropical rainforest. Encyclopædia Britannica, Inc.

larger South Pacific islands. Of the 13 angiosperm families generally recognized as the most primitive, all but two—Magnoliaceae and Winteraceae—are overwhelmingly tropical in their present distribution. Three families—Illiciaceae, Magnoliaceae, and Schisandraceae—are found predominantly in Northern Hemisphere rainforests. Five families—Amborellaceae, Austrobaileyaceae, Degeneriaceae, Eupomatiaceae, and Himantandraceae—are restricted to rainforests in the tropical Australasian region. Members of the Winteraceae are shared between this latter region and South America; those of the Lactoridaceae grow only on the southeast Pacific islands of Juan Fernández; members of the Canellaceae are shared between South America and Africa; and two families—Annonaceae and Myristicaceae—generally occur in tropical regions. This has led some authorities to suggest that the original cradle of angiosperm evolution might lie in Gondwanaland, a supercontinent of the Southern Hemisphere thought to have existed in the Mesozoic Era (251 to 65.5 million years ago) that consisted of Africa, South America, Australia, peninsular India, and Antarctica. An alternative explanation for this geographic pattern is that in the Southern Hemisphere, especially on islands, there are more refugia—that is, isolated areas whose climates remained unaltered while those of the surrounding areas changed, enabling archaic life-forms to persist.

 The first angiosperms are thought to have been massive, woody plants appropriate for a rainforest habitat. Most of the smaller, more delicate plants that are so widespread in the world today evolved later, ultimately from tropical rainforest ancestors. While it is possible that even earlier forms existed that await discovery, the oldest angiosperm fossils—leaves, wood, fruits, and flowers derived from trees—support the view that the earliest angiosperms were rainforest trees. Further evidence comes from the growth forms of the most primitive surviving angiosperms: all 13 of the most primitive angiosperm families consist of woody plants, most of which are large trees.

As the world climate cooled in the middle of the Cenozoic, it also became drier. This is because cooler temperatures led to a reduction in the rate of evaporation of water from, in particular, the surface of the oceans, which led in turn to less cloud formation and less precipitation. The entire hydrologic cycle slowed, and tropical rainforests—which depend on both warmth and consistently high rainfall—became increasingly restricted to equatorial latitudes. Within those regions rainforests were limited further to coastal and hilly areas where abundant rain still fell at all seasons. In the middle latitudes of both hemispheres, belts of atmospheric high pressure developed. Within these belts, especially in continental interiors, deserts formed. In regions lying between the wet tropics and the deserts, climatic zones developed in which rainfall adequate for luxuriant plant growth was experienced for only a part of the year. In these areas new plant forms evolved from tropical rainforest ancestors to cope with seasonally dry weather, forming tropical deciduous forests. In the drier and more fire-prone places, savannas and tropical grasslands developed.

Retreat of the rainforests was particularly rapid during the period beginning 5 million years ago leading up to and including the Pleistocene Ice Ages, or glacial intervals, that occurred between 2.6 million and 11,700 years ago. Climates fluctuated throughout this time, forcing vegetation in all parts of the world to repeatedly migrate, by seed dispersal, to reach areas of suitable climate. Not all plants were able to do this equally well because some had less-effective means of seed dispersal than others. Many extinctions resulted. During the most extreme periods (the glacial maxima, when climates were at their coldest and, in most places, also driest), the range of tropical rainforests shrank to its smallest extent, becoming restricted to relatively small refugia. Alternating intervals of climatic amelioration led to repeated range expansion, most recently from the close of the last glacial period about 10,000 years ago. Today large areas of tropical rainforest, such as Amazonia, have developed as a result of this relatively recent expansion. Within them it is possible to recognize "hot spots" of plant and animal diversity that have been interpreted as glacial refugia.

Tropical rainforests today represent a treasure trove of biological heritage. They not only retain many primitive plant and animal species but also are communities that exhibit unparalleled biodiversity and a great variety of ecological interactions. The tropical rainforest of Africa was the habitat in which the ancestors of humans evolved, and it is where the nearest surviving human relatives—chimpanzees and gorillas—live still. Tropical rainforests supplied a rich variety of food and other resources to indigenous peoples, who, for the most part, exploited this bounty without degrading the vegetation or reducing its range to any significant degree. However, in some regions a long history of forest burning by the inhabitants is thought to have caused extensive replacement of tropical rainforest and tropical deciduous forest with savanna.

Not until the past century, however, has widespread destruction of tropical forests occurred. Regrettably, tropical rainforests and tropical deciduous forests are now being destroyed at a rapid rate in order to provide resources such as timber and to create land that can be used for other purposes, such as cattle grazing. Today tropical forests, more than any other ecosystem, are experiencing habitat alteration and species extinction on a greater scale and at a more rapid pace than at any time in their history—at least since the major extinction event at the end of the Cretaceous Period, some 65.5 million years ago.

Tropical Rainforest Environments

The equatorial latitude of tropical rainforests and tropical deciduous forests keeps day length and mean temperature fairly constant throughout the year. The sun rises daily to a near-vertical position at noon, ensuring a high level of incoming radiant energy at all seasons. Although there is no cold season during which plants experience unfavourable temperatures that prohibit growth, there are many local variations in climate that result from topography, and these variations influence and restrict rainforest distribution within the tropics.

Tropical rainforests occur in regions of the tropics where temperatures are always high and where rainfall exceeds about 1,800 to 2,500 mm (about 70 to 100 inches) annually and occurs fairly evenly throughout the year. Similar hot climates in which annual rainfall lies between about 800 and 1,800 mm and in which a pronounced season of low rainfall occurs typically support tropical deciduous forests—that is, rainforests in which up to about three-quarters of the trees lose their leaves in the dry season. The principal determining climatic factor for the distribution of rainforests in lowland regions of the tropics, therefore, is rainfall, both the total amount and the seasonal variation. Soil, human disturbance, and other factors also can be important controlling influences.

The climate is always hot and wet in most parts of the equatorial belt, but in regions to its north and south seasonal rainfall is experienced. During the summer months of the Northern Hemisphere—June to August—weather systems shift northward, bringing rain to regions in the northern parts of the tropics, as do the monsoon rains of

India and Myanmar. Conversely, during the Southern Hemisphere's summer, weather systems move southward, bringing rain from December to February to places such as northern Australia. In these hot, seasonally wet areas grow tropical deciduous forests, such as the teak forests of Myanmar and Thailand. In other locations where conditions are similar but rainfall is not so reliable or burning has been a factor, savannas are found.

Topographic factors influence rainfall and consequently affect rainforest distribution within a region. For example, coastal regions where prevailing winds blow onshore are likely to have a wetter climate than coasts that experience primarily offshore winds. The west coasts of tropical Australia and South America south of the Equator experience offshore winds, and these dry regions can support rainforests only in very small areas. This contrasts with the more extensively rainforest-clad, east-facing coasts of these same continents at the same latitudes. The same phenomenon is apparent on a smaller scale where the orientation of coastlines is parallel to, rather than perpendicular to, wind direction. For example, in the Townsville area of northeastern Australia and in Benin in West Africa, gaps in otherwise fairly continuous tracts of tropical rainforest occur where the prevailing winds blow along the coast rather than across it.

Mean temperatures in tropical rainforest regions are between 20 and 29 °C (68 and 84 °F), and in no month is the mean temperature below 18 °C (64 °F). Temperatures become critical with increasing altitude; in the wet tropics temperatures fall by about 0.5 °C (0.9 °F) for every 100 metres (328 feet) climbed. Vegetation change across altitudinal gradients tends to be gradual and variable and is interpreted variously by different authorities. For example, in Uganda tropical rainforest grows to an altitude of 1,100 to 1,300 metres (3,600 to 4,300 feet) and has been described as giving way, via a transition forest zone, to montane rainforest above 1,650 to 1,750 metres (5,400 to 5,700 feet), which continues to 2,300 to 3,400 metres (7,500 to 11,200 feet). In New Guinea, lowland tropical rainforest reaches 1,000 to 1,200 metres (3,300 to 3,900 feet), above which montane rainforests extend, with altitudinal variation, to 3,900 metres (12,800 feet). In Peru, lowland rainforest extends upward to 1,200 to 1,500 metres (3,900 to 4,900 feet), with transitional forest giving way to montane rainforest above 1,800 to 2,000 metres (5,900 to 6,600 feet), which continues to 3,400 to 4,000 metres (11,200 to 13,100 feet). These limits are comparable and reflect the similarities of climate in all regions where tropical rainforests occur. Plant species, however, are often quite different among regions.

Although the climate supporting tropical rainforests is perpetually hot, temperatures never reach the high values regularly recorded in drier places to the north and south of the equatorial belt. This is partly due to high levels of cloud cover, which limit the mean number of sunshine hours per day to between four and six. In hilly areas where air masses rise and cool because of the topography, the hours of sunlight may be even fewer. Nevertheless, the heat may seem extreme owing to the high levels of atmospheric humidity, which usually exceed 50 percent by day and approach 100 percent at night. Exacerbating the discomfort is the fact that winds are usually light; mean wind speeds are generally less than 10 km (6.2 miles) per hour and less than 5 km (3.1 miles) per hour in many areas. Devastating hurricanes (cyclones and typhoons) occur periodically in some coastal regions toward the margins of the equatorial belt, such as in the West Indies and in parts of the western Pacific region. Although relatively infrequent, such storms have an important effect on forest structure and regeneration.

The climate within any vegetation (microclimate) is moderated by the presence of plant parts that reduce incoming solar radiation and circulation of air. This is particularly true in tropical rainforests, which are structurally more dense and complex than other vegetation. Within the forest, temperature range and wind speed are reduced and humidity is increased relative to the climate above the tree canopy or in nearby clearings. The amount of rain reaching the ground is also reduced—by as much as 90 percent in some cases—as rainwater is absorbed by epiphytes (plants that grow on the surface of other plants but that derive nutrients and water from the air) and by tree bark or is caught by foliage and evaporates directly back to the atmosphere.

Soils in tropical rainforests are typically deep but not very fertile, partly because large proportions of some mineral nutrients are bound up at any one time within the vegetation itself rather than free in the soil. The moist, hot

Epiphytic orchids (Dendrobium). E.R. Degginger

climatic conditions lead to deep weathering of rock and the development of deep, typically reddish soil profiles rich in insoluble sesquioxides of iron and aluminum, commonly referred to as tropical red earths. Because precipitation in tropical rainforest regions exceeds evapotranspiration at almost all times, a nearly permanent surplus of water exists in the soil and moves downward through the soil into streams and rivers in valley floors. Through this process nutrients are leached out of the soil, leaving it relatively infertile. Most roots, including those of trees, are concentrated in the uppermost soil layers where nutrients become available from the decomposition of fallen dead leaves and other organic litter. Sandy soils, particularly, become thoroughly leached of nutrients and support stunted rainforests of peculiar composition. A high proportion of plants in this environment have small leaves that contain high levels of toxic or unpalatable substances. A variant of the tropical rainforest, the mangrove forest, is found along estuaries and on sheltered sea coasts in tidally inundated, muddy soils.

Even within the same area, however, there are likely to be significant variations in soil related to topographic position and to bedrock differences, and these variations are reflected in forest composition and structure. For example, as altitude increases—even within the same area and on the same bedrock—soil depth decreases markedly and its organic content increases in association with changes in forest composition and structure.

The Biota of Tropical Rainforests

Only a minority of plant and animal species in tropical rainforests and tropical deciduous forests have been described formally and named. Therefore, only a rough estimate can be given of the total number of species contained in these ecosystems, as well as the number that are becoming extinct as a result of forest clearance. Nevertheless, it is quite clear that these vegetation types are the most diverse of all, containing more species than any other ecosystem. This is particularly so in regions in which tropical rainforests not only are widespread but also are separated into many small areas by geographic barriers, as in the island-studded Indonesian region. In this area different but related species often are found throughout various groups of islands, adding to the total regional diversity. Exceptionally large numbers of species also occur in areas of diverse habitat, such as in topographically or geologically complex regions and in places that are believed to have acted as refugia throughout the climatic fluctuations of the past few million years. According to some informed estimates, more than a hundred species of rainforest fauna and flora become extinct every week as a result of widespread clearing of forests by humans. Insects are believed to constitute the greatest percentage of disappearing species.

Flora

All major groups of terrestrial organisms are represented abundantly in tropical rainforests. Among the higher plants, angiosperms are particularly diverse and include many primitive forms and many families not found in the vegetation of other ecosystem types. Many flowering plants are large trees, of which there is an unparalleled diversity. For example, in one area of 23 hectares (57 acres) in Malaysia, 375 different tree species with trunk diameters greater than 91 cm (35.8 inches) have been recorded, and in a 50-hectare (124-acre) area in Panama, 7,614 trees belonging to 186 species had trunk diameters greater than 20 cm (7.8 inches). New species of plants—even those as conspicuously large as trees—are found every year. Relatively few

FORESTS AND GRASSLANDS

THE MONSOON FOREST

Monsoon forests, which are also known as dry forests or tropical deciduous forests, are open woodland in tropical areas that have a long dry season followed by a season of heavy rainfall. The trees in a monsoon forest usually shed their leaves during the dry season and come into leaf at the start of the rainy season. Many lianas (woody vines) and herbaceous epiphytes (air plants, such as orchids) are present. Monsoon forests are especially well developed in Southeast Asia and are typified by tall teak trees and thickets of bamboo.

Monsoon forest in the Anaimalai Hills, Western Ghats, Tamil Nadu state, India. Tropical deciduous forests grow in seasonally dry areas of the tropics. These forests occur mainly in India, the Myanmar–Vietnam–southern coastal China region, eastern Brazil, smaller areas in South and Central America north of the Equator, the West Indies, southeastern Africa, and northern Australia. Gerald Cubitt

gymnosperms (conifers and their relatives), however, are found in rainforests; instead, they occur more frequently at the drier and cooler extremes of the range of climates in which tropical rainforests grow. Some plant families, such as Arecaceae (palms), are typically abundant in all tropical rainforest regions, although different species occur from region to region. Other families are more restricted geographically. The family Dipterocarpaceae (dipterocarps) includes many massive trees that are among the most abundant and valuable species in the majority of tropical rainforests in western Malesia; the family, however, is uncommon in New Guinea and Africa and absent from South and Central America and Australia. The Bromeliaceae (bromeliads), a large family consisting mainly of rainforest epiphytes and to which the pineapple belongs, is entirely restricted to the New World.

Tropical rainforests, which contain many different types of trees, seldom are dominated by a single species. A species can predominate, however, if particular soil conditions favour this occurrence or minimal disturbance occurs for several tree generations. Tropical deciduous forests are less diverse and often are dominated by only one or two tree species. The extensive deciduous forests of Myanmar, for example, cover wide areas and are dominated by only one or two tree species—teak (*Tectona grandis*) and the smaller leguminous tree *Xylia xylocarpa*. In Thailand and Indochina deciduous forests are dominated by members of the Dipterocarpaceae family, *Dipterocarpus tuberculatus*, *Pentacme suavis*, and *Shorea obtusa*.

Ferns, mosses, liverworts, lichens, and algae are also abundant and diverse, although not as well studied and cataloged as the higher plants. Many are epiphytic and are found attached to the stems and sometimes the leaves of larger plants, especially in the wettest and most humid

LIFE IN A BROMELIAD POOL

Bromeliads comprise an entire order of flowering plants called Bromeliales. The pineapple is the most familiar member of this tropical American group, which also includes some of the most interesting plants of the rainforest—the tank bromeliads. Most bromeliads are epiphytes—that is, plants that live attached to other vegetation. Many live high above the forest floor, deriving energy from photosynthesis, water from rain, and nutrients mainly from falling debris and windblown dust.

The tank bromeliads have relationships with a wide variety of other organisms. The water held in the leaf rosette of a tank bromeliad forms a virtual aquarium, which may contain up to 20 litres (5 gallons) of water. Several hundred species of aquatic organisms can be found in these habitats, and some are found nowhere else except in bromeliad pools. Among the creatures found here are fungi, algae, protozoa, and small invertebrates such as insects, spiders, scorpions, mites, worms, and even crabs. Vertebrate inhabitants of bromeliad tanks include frogs, salamanders, and snakes. Animal life, however, is dominated by insects, especially dipterans (two-winged flies) such as nonbiting midges and mosquitoes. On occasion, an aquatic species of bladderwort can be found floating in bromeliad tanks.

These small, discrete, relatively stable communities can serve as valuable models for studying biological processes. In a typical food web of a bromeliad pool, energy and nutrients flow from solutes and organic detritus in the water, through bacteria and protozoa, to browsing or filter-feeding mosquito larvae, and thence to aquatic predators such as crabs, larvae of other mosquitoes, and damselflies. Within the confines of a pool the risk of predation is severe. Among the predators are two genera of damselfly (*Diceratobasis* and *Leptagrion*) that are known from no other habitat. Predator becomes prey, however, should a bromeliad crab (*Metopaulias depressus*) choose the pool for its offspring. In order to protect its larvae from such predators, the crab kills all damselfly larvae in a pool before placing its own progeny there.

A female strawberry poison frog (*Dendrobates pumilio*) uses a different strategy to protect her young. She transports one or two newly hatched tadpoles from the leaves on which her eggs are laid to a bromeliad pool, which serves as a nursery. She then exhibits parental care by depositing in the pool nutritive (nonviable) eggs on which the developing tadpoles feed.

Tree frog tadpoles, on the other hand, must fend for themselves within the pool. Mosquito larvae are commonly fed upon by the tadpoles, and certain larvae are not safe even from each other. *Toxorhynchites* mosquito larvae are both predatory and cannibalistic, and individuals are especially vulnerable to cannibalism just after molting. In a preemptive strategy a large larva, about to become a pupa, will doggedly kill, but not consume, any other mosquito larva it encounters.

The *Anopheles* mosquito, the vector for the organism that causes malaria in humans, requires fresh standing water in order to complete the larval stages of its life cycle. When it was discovered that tank bromeliads are ideal sites for the mosquito to complete its life cycle, programs of bromeliad eradication were implemented as one part of the overall effort to eliminate *Anopheles* from malaria-plagued regions.

places. Fungi and other saprophytic plants (vegetation growing on dead or decaying matter) are similarly diverse. Some perform a vital role in decomposing dead organic matter on the forest floor and thereby releasing mineral nutrients, which then become available to roots in the surface layers of the soil. Other fungi enter into symbiotic relationships with tree roots (mycorrhizae).

Fauna

Interacting with and dependent upon this vast array of plants are similarly numerous animals. Like the plants, most animal species are limited to only one or a few types of tropical rainforest within an area, with the result that the overall number of species is substantially greater than it is in a single forest type. For example, a study of insects in the canopy of four different types of tropical rainforest in Brazil revealed 1,080 species of beetle, of which 83 percent were found in only one forest type, 14 percent in two, and only 3 percent in three or four types. While the larger, more conspicuous vertebrates (mammals, birds, and to a lesser degree amphibians and reptiles) are well known, only a small minority of the far more diverse invertebrates (particularly insects) have ever been collected, let alone described and named.

As with the plants, some animal groups occur in all tropical rainforest regions. A variety of fruit-eating parrots, pigeons, and seed-eating weevil beetles, for example, can be expected to occur in any tropical rainforest. Other groups are more restricted. Monkeys, while typical of tropical rainforests in both the New and the Old World, are entirely absent from New Guinea and areas to its east and south. Tree kangaroos inhabit tropical rainforest canopies only in Australia and New Guinea, and birds of paradise are restricted to the same areas.

To a large extent these geographic variations in tropical rainforest biota reflect the long-term geologic histories of these ancient ecosystems. This is most clearly demonstrated in the Malesian phytogeographic subkingdom, which has existed as a single entity only since continental movements brought Australia and New Guinea northward into juxtaposition with Southeast Asia about 15 million years ago. Before that time the two parts were separated by a wide expanse of ocean and experienced separate evolution of their biota. Only a relatively small sea gap lies between them today; Java, Bali, and Borneo are on one side, and Timor and New Guinea are on the other, with islands such as Celebes and the Moluccas forming an intermediate region between. The gap is marked by a change in flora and, especially, fauna and is known as Wallace's Line. The contrast is particularly stark with respect to mammals. To the west the rainforests are populated—or were populated until recently—by monkeys, deer, pigs, cats, elephants, and rhinoceroses, while those to the east have marsupial mammals, including opossums, cuscuses, dasyurids, tree kangaroos, and bandicoots. Only a few groups such as bats and rodents have migrated across the line to become common in both areas. Similar contrasts,

THE CLOUD FOREST

A cloud forest, or montane rainforest, is made up of the vegetation of tropical mountainous regions in which the rainfall is often heavy and persistent condensation occurs because of cooling of moisture-laden air currents deflected upward by the mountains. The trees in a cloud forest are typically short and crooked. Mosses, climbing ferns, lichens, and epiphytes (air plants, such as orchids) form thick blankets on the trunks and branches of the trees. Begonias, ferns, and many other herbaceous plants may grow to exceptionally large size in clearings. A forest of extremely stunted, moss-covered trees that occurs in tropical or temperate mountainous regions is sometimes known as an elfin woodland.

albeit less pronounced, can be seen in many other animal and plant groups across the same divide.

POPULATION AND COMMUNITY DEVELOPMENT AND STRUCTURE OF TROPICAL RAINFORESTS

Tropical rainforests are distinguished not only by a remarkable richness of biota but also by the complexity of the interrelationships of all the plant and animal inhabitants that have been evolving together throughout many millions of years. As in all ecosystems, but particularly in the complex tropical rainforest community, the removal of one species threatens the survival of others with which it interacts. Some interactions are mentioned below, but many have yet to be revealed.

The General Structure of the Rainforest

Plants with similar stature and life-form can be grouped into categories called synusiae, which make up distinct layers of vegetation. In tropical rainforests the synusiae are more numerous than in other ecosystem types. They include not only mechanically independent forms, whose stems are self-supporting, and saprophytic plants but also mechanically dependent synusiae such as climbers, stranglers, epiphytes, and parasitic plants. An unusual mix of trees of different sizes is found in the tropical rainforest, and those trees form several canopies below the uppermost layer, although they are not always recognizably separate layers. The upper canopy of the tropical rainforest is typically greater than 40 metres (131 feet) above ground.

The tropical rainforest is structurally very complex. Its varied vegetation illustrates the intense competition for light that goes on in this environment in which other climatic factors are not limiting at any time of year and the vegetation is thus allowed to achieve an unequaled luxuriance and biomass. The amount of sunlight filtering through the many layers of foliage in a tropical rainforest is small; only about 1 percent of the light received at the top of the canopy reaches the ground. Most plants depend on light for their energy requirements, converting it into chemical energy in the form of carbohydrates by the process of photosynthesis in their chlorophyll-containing green tissues. Few plants can persist in the gloomy environment at ground level, and the surface is marked by a layer of rapidly decomposing dead leaves rather than of small herbaceous plants. Mosses grow on tree butts, and there are a few forbs such as ferns and gingers, but generally the ground is bare of living plants, and even shrubs are rare. However, tree seedlings and saplings are abundant; their straight stems reach toward the light but receive too little energy to grow tall enough before food reserves from their seeds are exhausted. Their chance to grow into maturity comes only if overhanging vegetation is at least partially removed through tree death or damage by wind. Such an occurrence permits more solar radiation to reach their level and initiates rapid growth and competition between saplings as to which will become a part of the well-lit canopy.

"FLYING" TREES: AERIAL SEED DISPERSAL IN THE PANAMANIAN RAINFOREST

As in most tropical forests, the trees of Panama exhibit a variety of different adaptations to aid dispersal of their seeds. These adaptations involve substantial investment of the trees' material, but they are worthwhile because seed dispersal increases both the seeds' and the species' chances of survival. Seed destroyers such as herbivores, fungi, and bacteria often concentrate their activities in the vicinity of the parent tree. Therefore, seeds that can come to rest some distance away from the parent tree are more likely to germinate and grow.

Dispersal efforts that take advantage of air currents can be elaborate. Because the rainforest canopy effectively blocks wind from reaching the environment below, aerial seed dispersal is not as widely afforded as in other, more open ecosystems. Even so, many trees have managed to exploit this strategy. For example, the kapok tree, found in tropical forests throughout the world, is an emergent—a tree whose crown rises well above the canopy. The kapok's towering height enables it to gain access to winds above the canopy. The tiny seeds of the kapok are attached to fine fibres that, when caught by the wind, enable distribution far from the parent tree. The balsa tree also uses fibrous seeds to distribute its progeny, but it is not an emergent. Instead, balsa grows quickly as a colonizer of gaps in the forest, giving its seeds access to wind while the gap in the trees is still open.

Other trees grow aerodynamic structures to make use of the wind. The canopy trees *Platypodium elegans* and *Tachigalia versicolor* produce single-winged fruits similar to those of maple

Woolly seeds produced by the seed pods of the kapok tree (Ceiba pentandra). Norman Myers—Bruce Coleman Inc.

trees common in temperate zones. In the case of *P. elegans*, each fruit is attached to a twig by the tip of its wing and has a dry weight of about 2 grams (0.07 ounce)—only about 20 percent of which is the seed's weight. They remain unripe for many months, but when Panama's dry season arrives (January through March) the fruits dry out and are dispersed by strong seasonal winds. Seeds often are blown 50 metres (160 feet) or more. Shaded seedlings within about 30 metres (100 feet) of the parent tree tend to die from fungal attack, but fruits landing farther than 30 metres from the tree or in canopy gaps fare much better. The suicide tree encloses its seeds in elliptical wings that can measure nearly 15 cm (6 inches) long. The tree's name comes from the fact that, after producing seeds, the tree dies.

Gaps in the canopy of a tropical rainforest provide temporarily well-illuminated places at ground level and are vital to the regeneration of most of the forest's constituent plants. Few plants in the forest can successfully regenerate in the deep shade of an unbroken canopy; many tree species are represented there only as a population of slender, slow-growing seedlings or saplings that have no chance of growing to the well-lit canopy unless a gap forms. Other species are present, invisibly, as dormant seeds in the soil. When a gap is created, seedlings and saplings accelerate their growth in the increased light and are joined by new seedlings sprouting from seeds stored in the soil that have been stimulated to germinate by light or by temperature fluctuations resulting from the sun's shining directly on the soil surface. Other seeds arrive by various seed-dispersal processes. A thicket of regrowth rapidly develops, with the fastest-growing shrubs and trees quickly shading out opportunistic, light-demanding, low-growing herbaceous plants and becoming festooned with lianas. Through it all slower-growing, more shade-tolerant but longer-lived trees eventually emerge and restore the full forest canopy. The trees that initially fill in the gap in the canopy live approximately one century, whereas the slower-growing trees that ultimately replace them may live for 200 to 500 years or, in extreme cases, even longer. Detailed mapping of the trees in a tropical rainforest can reveal the locations of previous gaps through identification of clumps of the quicker-growing, more light-demanding species, which have yet to be replaced by trees in the final stage of successional recovery. Local, natural disturbances of this sort are vital to the maintenance of the full biotic diversity of the tropical rainforest.

Just as tropical rainforest plants compete intensely for light above ground, below ground they vie for mineral nutrients. The process of decomposition of dead materials is of crucial importance to the continued health of the forest because plants depend on rapid recycling of mineral nutrients. Bacteria and fungi are primarily responsible for this process. Some saprophytic flowering plants that occur

An earthstar (Geastrum) *puffball, growing on moist soil among mosses.* Larry West—The National Audubon Society/Photo Researchers

in tropical rainforests rely on decomposing material for their energy requirements and in the process use and later release minerals. Some animals are important in the decomposition process; for example, in Malaysia termites have been shown to be responsible for the decomposition of as much as 16 percent of all litter, particularly wood. Most trees in the tropical rainforest form symbiotic mycorrhizal associations with fungi that grow in intimate contact with their roots; the fungi obtain energy from the tree and in turn provide the tree with phosphorus and other nutrients, which they absorb from the soil very efficiently. A mat of plant roots explores the humus beneath the rapidly decomposing surface layer of dead leaves and twigs, and even rotting logs are invaded by roots from below. Because nutrients are typically scarce at depth but, along with moisture, are readily available in surface layers, few roots penetrate very deeply into the soil. This shallow rooting pattern increases the likelihood of tree falls during storms, despite the support that many trees receive from flangelike plank buttresses growing radially outward from their trunk bases. When large trees fall, they may take with them other trees against which they collapse or to which they are tied by a web of lianas and thereby create gaps in the canopy.

Tree growth requires substantial energy investment in trunk development, which some plants avoid by depending on the stems of other plants for support. Perhaps the most obvious adaptation of this sort is seen in plants that climb from the ground to the uppermost canopy along other plants by using devices that resemble grapnel-like hooks. Lianas are climbers that are abundant and diverse in tropical rainforests; they are massive woody plants whose mature stems often loop through hundreds of metres of forest, sending shoots into new tree crowns as successive supporting trees die and decay. Climbing palms

FORESTS

Lianas in a tropical rainforest. The vascular tissues of lianas are modified primarily for water conduction, which leaves these tall plants dependent on other plants for support. © Gary Braasch

or rattans (*Calamus*) are prominent lianas in Asian rainforests, where the stems, which are used to make cane furniture, provide a valuable economic resource.

Epiphytes are particularly diverse and include large plants such as orchids, aroids, bromeliads, and ferns in addition to smaller plants such as algae, mosses, and lichens. In tropical rainforests epiphytes are often so abundant that their weight fells trees. Epiphytes that grow near the upper canopy of the forest have access to bright sunlight but must survive without root contact with the soil. They depend on rain washing over them to provide water and mineral nutrients. During periods of drought, epiphytes undergo stress as water stored within their tissues becomes depleted. The diversity of epiphytes in tropical deciduous forests is much less than that of tropical rainforests because of the annual dry season.

Parasitic flowering plants also occur. Hemiparasitic mistletoes attached to tree branches extract water and minerals from their hosts but carry out their own photosynthesis. Plants that are completely parasitic also are found in tropical rainforests. *Rafflesia*, in Southeast Asia, parasitizes the roots of certain lianas and produces no aboveground parts until it flowers; its large orange and yellow blooms, nearly 1 metre (3.28 feet) in diameter, are the largest flowers of any plant.

Stranglers make up a type of synusia virtually restricted to tropical rainforests. In this group are strangler figs (*Ficus*), which begin life as epiphytes, growing from seeds left on high tree branches by birds or fruit bats. As they grow, they develop long roots that descend along the trunk of the host tree, eventually reaching the ground and entering the soil. Several roots usually do this, and they become grafted together as they crisscross each other to form a lattice, ultimately creating a nearly complete sheath around the trunk. The host tree's canopy becomes shaded by the thick fig foliage, its trunk constricted by the surrounding root sheath and its own root system forced to compete with that of the strangling fig. The host tree is also much older than the strangler and eventually dies and rots away, leaving a giant fig "tree" whose apparent "trunk" is actually a cylinder of roots, full of large hollows that provide shelter and breeding sites for bats, birds, and other animals. Stranglers may also develop roots from their branches, which, when they touch the ground, grow into the soil, thicken, and become additional "trunks." In this way stranglers grow outward to become large patches of fig forest that consist of a single plant with many interconnected trunks.

The Relationships Between the Flora and Fauna

Some of the tallest trees and lianas, and the epiphytes they support, bear flowers and fruits at the top of the rainforest canopy, where the air moves unfettered by vegetation. They are able to depend on the wind for dispersal of pollen from flower to flower, as well as for the spreading of fruits and seeds away from the immediate environment of the parent plant. Ferns, mosses, and other lower plants also exploit the wind to carry their minute spores. However, a great many flowering plants, including many that grow in the nearly windless environment of the understory, depend on animals to perform these functions. They are as dependent on animals for reproductive success as the animals are on them for food—one example of the mutual dependence between plants and animals.

Many rainforest trees have sizable seeds from which large seedlings emerge and thrust their way through the thick mat of dead leaves on the dark forest floor. They develop tall stems, using food reserves in the seed without having to rely on sunlight, which is usually too dim, to meet their energy requirements. Because large seeds cannot be dispersed by the wind, these plants depend on a

RAINFOREST REGENERATION IN PANAMA

Forest regeneration, following such events as forest clearing by humans or as part of a natural process, results from interactions among diverse groups of organisms and the environment. Depending upon factors such as survivorship, pollination, and seed production and dispersal, different tree species will be represented. Physical factors that can limit plant growth by blocking access to light, water, and nutrients strongly influence the outcome of regeneration. For example, most tree species require openings in the forest canopy (canopy gaps) in order to receive sufficient light to attain a mature size and stature, but the seedlings of different tree species show very different requirements for light. Tropical forest tree species in Panama tend to assort along a continuum of characteristics that relate to how they grow and reproduce. This continuum can be thought of as a series of trade-offs. At one extreme are fast-growing pioneer species such as balsa or cecropia. These trees are characterized by rapid growth in high light, high mortality (especially in shaded environments), low wood densities, and relatively rapid attainment of reproductive status. They also tend to produce leaves with high photosynthetic capacities that flush green but suffer high levels of insect damage, consequently lowering the trees' lifetimes. At the other extreme are tree species such as *Manilkara*, almendro, and the suicide tree, characterized by slower growth and lower light requirements, with the capacity for extended persistence under low light conditions. Such trees tend toward high wood densities, relatively delayed attainment of reproductive status, and larger, often animal-dispersed seeds. They also have tough, long-lived, frequently reddish leaves that exhibit relatively low photosynthetic rates. These differences in characteristics associated with different life histories reflect the various ways that plants have evolved to deal with the complexities of living in a tropical forest.

variety of animals to perform this function and have evolved many adaptations to encourage them to do so. Fruit bats are attracted by fragrant, sweet fruits typically borne conspicuously and conveniently on the outer parts of the tree canopy; the mango (*Mangifera indica*), native to the rainforests of India, provides a good example. The bats not only feed on fruits as they hang from the trees but also may carry a fruit away to another perch, where they

APARTMENTS OF THE RAINFOREST: COMMUNITIES IN TREE HOLLOWS

Tree hollows are sought-after refuges for a succession of creatures, from termites to primates. Tree hollows make safe nests and dens where mothers can raise their young protected from predators and where roosting birds and various mammals can take shelter during the day.

The creation of a tree hollow involves a complex interaction between the largest and longest-lived creatures on the planet—trees—and some of the smallest—bacteria—and then continues with the participation of hundreds of other vertebrate and invertebrate intermediaries. The process is initiated by an injury to the tree, often from insects, wind, birds, people, or bark-damaging species such as marmosets. Trees cannot heal damaged tissue; instead they respond to injury by erecting physical and chemical barriers to separate healthy from damaged tissue and thus prevent bacteria and fungi from colonizing their water conduits. As microorganisms break down damaged tree tissue, the tree attempts to wall off, or compartmentalize, the wound. Because of compartmentalization, a tree may continue to survive with a hollow cylinder at its core—a phenomenon that is particularly common in the American tropics.

Many tropical bats, such as ghost bats, bulldog bats, and vampire bats, prefer hollow trees as day roosts, where scorpions, centipedes, roaches, and termites are attracted to their guano. When the bats leave to hunt flying insects, mammals such as opossums visit the hollows to hunt the guano-loving invertebrates.

Various birds, including piculets and the majority of the world's tropical parrots, chisel out nest cavities from trees that are already significantly rotted; others, such as tityras, take over cavities excavated by previous tenants, just as nesting toucans and toucanets take over tree-hollow nests vacated by woodpeckers in South American rainforests. Tree frogs and lizards also take refuge in old woodpecker excavations.

eat the flesh and drop the seed. Smaller fruits may be swallowed whole, the seeds passing through the gut intact and being voided at a distance. The ground beneath trees used by fruit bats as a roost is commonly thick with seedlings of fleshy, fruit-bearing trees.

A variety of birds eat fleshy fruits also, voiding or regurgitating the unharmed seeds. Birds of different sizes are typically attracted to similarly scaled fruits, which are carried on stems of appropriate thickness and strength. For example, large pigeons in New Guinea feed preferentially on larger fruits borne on thicker stems that can bear not only the weight of the fruit but also the weight of the large bird; smaller pigeons tend to feed on smaller fruits borne on thinner twigs. In such a manner, the diverse plant community is matched by a similarly diverse animal community in interdependence.

Terrestrial mammals also help to disperse seeds. In many cases this has favoured the positioning of flowers and fruits beneath the canopy on the trunks of trees accessible to animals unable to climb or fly, an adaptation called cauliflory. In some cases fruits are grown in the canopy but drop as they ripen, opening only after they fall to attract ground-dwelling animals that will carry them away from the parent tree. The durian fruit (*Durio zibethinus*) of Southeast Asian rainforests is an example; its fruits are eaten and its seeds dispersed by a range of mammals, including pigs, elephants, and even tigers.

Mango (Mangifera indica). Robert C. Hermes from the National Audubon Society Collection/Photo Researchers—EB Inc.

HITCHING A RIDE: SEED DISPERSAL BY ANIMALS IN THE PANAMANIAN RAINFOREST

Numerous plants depend on animal dispersers to transport seeds either internally or externally. Birds generally disperse seeds internally by eating the fruits, which are often small and red and the numerous seeds of which easily pass through the birds' digestive systems. Some seeds actually have higher rates of germination after passing through animal gut; others benefit from being deposited in nutrient-rich dung. Fruit bats such as the Jamaican, or common, fruit bat (*Artibeus jamaicensis*) are important seed dispersers in Panama, feeding on many fruits, including those of figs (genus *Ficus*) and cecropias (genus *Cecropia*), and distributing some seeds internally and others externally. The bat homes in on the smell of ripe fruit and transports it to a feeding roost away from the source tree. Small seeds are eaten and later excreted in flight, whereas larger seeds are discarded at the feeding site.

*Barbados cherry (*Malpighia glabra*). Douglas David Dawn*

Other examples of external seed transport by animals are also common. Some trees provide rich fruit that is attractive to foraging animals. As a consequence, organisms ranging from ants to bats to rodents such as the agouti unwittingly disperse the trees' seeds. For example, the wild cashew (*Anacardium excelsum*) bears nuts on a sweet, green stem enlargement (hypocarp) that is a favourite food of many bats, which disperse the nuts while feeding.

The seed dispersal process can be complex, involving the activity of more than one animal, or it may depend on specific animal behaviours. The bright orange fruits of the black palm (*Astrocaryum standleyanum*), for example, comprise a seed covered by a tough woody layer forming a nut, or stone, which is in turn covered by a layer of pulp. When the fruit ripens and drops to the forest floor, many animals come to eat the sweet pulp, sometimes moving the seeds about in the process. Since weevils lay eggs on nearly all black palm fruits, unless agoutis peel the orange flesh from the palm nuts and bury them, the newly hatched weevil larvae destroy the seeds. Therefore, despite the fact that they eat large numbers of the seeds themselves, agoutis provide a net benefit to the palm. In the absence of agoutis it is likely that a tract of forest with *Astrocaryum* would offer few prospects for new trees.

Agoutis are also important to the almendro tree (*Dipteryx panamensis*), which attracts many dispersers because it fruits at the end of Panama's dry season, when fruit is in short supply. A single seed is encased in a thick, hard wooden pod covered with a thin layer of green pulp. When a fruit crop ripens, numerous arboreal animals flock to it, including kinkajous, bats, monkeys, coatis, and squirrels. In addition, ground dwellers such as agoutis, peccaries, pacas, spiny rats, and tapirs seek out fruits that fall to the forest floor. Most of these animals simply eat the sweet pulp covering the fruit, but for the almendro seed to germinate it must first be carried far from its parent tree and buried. In the case of the almendro, the process is initiated by 70-gram (2.5-ounce) fruit bats (*Artibeus lituratus*), which first disperse a large number of fruits by carrying them off to feeding roosts away from the parent tree, where they chew off the pulp and drop the seeds. Then agoutis, which are less likely to bury almendro seeds found near parent trees, carry off seeds that the bats have dropped and bury some of them. Normally, agoutis consume most of these seeds or eat the seedlings when they germinate, but in a year of abundant fruit buried seeds will often germinate and grow. Thus, the almendro may need two animals, the fruit bat and the agouti, to give its seeds the opportunity to become new trees. Such findings strongly suggest that, in order to conserve many of the tree species in a tropical forest, it is also important to protect animal populations.

Cashew apples (hypocarp) and nuts of the domesticated cashew tree (Anacardium occidentale). W.H. Hodge

Many other animals, from ants to apes, are involved in seed dispersal. In the Amazon basin of Brazil, where large areas of tropical rainforest are seasonally flooded, many trees produce fruit attractive to fish, which swallow them whole and void the seeds. Squirrels are also important seed dispersers in parts of South America. In the tropical rainforests of northeastern Australia, cassowaries are responsible for generating mixed clumps of tree seedlings of several species that grow from their dung sites.

It is important for seeds to be spread away from parent plants, both to allow seedlings to escape competition with the parent and to expand the range of the species. Another capacity important to seed survival, particularly in the diverse tropical rainforest community, involves the evasion of seed predators. Many different beetles and other insects are specialized to feed on particular types of seed. Seeds concentrated beneath a parent plant are easy for seed predators to locate. Seeds that are carried away to areas occupied by different plant species—and different seed predators—are more likely to survive.

In addition to dispersing seeds, animals are vital to tropical rainforest reproduction through flower pollination. Many insects such as bees, moths, flies, and beetles as well as birds and bats carry out this activity. Birds such as the hummingbirds of South and Central America and the flower-peckers of Asia have adaptations that allow them to sip nectar from flowers. In the process they inadvertently become dusted with pollen, which they subsequently transport to other flowers, pollinating them. The plants involved also show special adaptations in flower structure and colour. Most flowers pollinated by birds are red, a colour highly visible to these animals, whereas flowers pollinated by night-flying moths are white or pink, and those pollinated by insects that fly during the day are often yellow or orange. Bats are important pollinators of certain pale, fragrant flowers that open in the evening in Asian rainforests.

When two or more species in an ecosystem interact to each other's benefit, the relationship is said to be mutualistic. The production of Brazil nuts and the regeneration of the trees that produce them provide an example of mutualism, and in this case the interaction also illustrates the importance of plant and animal ecology in maintaining a rainforest ecosystem.

Euglossine bees (most often the females) are the only creatures regularly able to gain entrance to the Brazil nut tree's flowers, which have lids on them. The bees enter to feed on nectar, and in the process they pollinate the flower. Pollination is necessary to initiate the production of nuts by the tree. Thus, the Brazil nut tree depends on female euglossine bees for pollination.

Male euglossines have a different role in this ecological process. To reproduce, the males must first prove themselves to the females. The males accomplish this by visiting orchids for the single purpose of gathering fragrant chemicals from the flowers. These fragrances are a necessary precondition of euglossine mating. Without the orchids of the surrounding rainforest, the euglossine population cannot sustain itself, and the Brazil nut trees do not get pollinated. For this reason, Brazil nuts used for human consumption must be collected from the rainforest; they cannot be produced on plantations.

Once the Brazil nut pods are formed, the tree then depends on the agouti, a rodent, to distribute and actually plant the seeds. The agouti is one of the few animals capable of chewing through the very hard pod to reach the nuts inside. Agoutis scatter and bury the nuts for future consumption, but some nuts manage to sprout and grow into mature trees.

The Biological Productivity of Tropical Rainforests

Of all vegetation types, tropical rainforests grow in climatic conditions that are least limiting to plant growth. It is to be expected that the growth and productivity (total amount of organic matter produced per unit area per unit time) of tropical rainforests would be higher than that of other vegetation, provided that other factors such as soil fertility or consumption by herbivorous animals are not extremely low or high.

Various methods are employed to assess productivity. Gross primary productivity is the amount of carbon fixed during photosynthesis by all producers in the ecosystem. However, a large part of the harnessed energy is used up by the metabolic processes of the producers (respiration). The amount of fixed carbon not used by plants is called net primary productivity, and it is this remainder that is available to various consumers in the ecosystem—such as the herbivores, decomposers, and carnivores. Of course, in any stable ecosystem there is neither an accumulation nor a diminution in the total amount of organic matter present, so that overall there is a balance between the gross primary productivity and the total consumption. The amount of organic matter in the system at any point in time, the total mass of all the organisms present, is called the biomass.

The biomass of tropical rainforests is larger than that of other vegetation. It is not an easy quantity to measure, involving the destructive sampling of all the plants in an area (including their underground parts), with estimates made of the mass of other organisms belonging to the ecosystem such as animals. Measurements show that tropical rainforests typically have biomass values on the order of 400 to 700 metric tons per hectare, greater than most temperate forests and substantially more than other vegetation with fewer or no trees. A measurement of biomass in a tropical deciduous forest in Thailand yielded a value of about 340 metric tons per hectare.

Increase in biomass over the period of a year at one rainforest site in Malaysia was estimated at 7 metric tons per hectare, while total litter fall was 14 metric tons, estimated mass of sloughed roots was 4 metric tons, and total live plant matter eaten by herbivorous animals (both invertebrate and vertebrate) was about 5 metric tons per hectare per year. These values add up to a total net production of 30 metric tons per hectare per year. Respiration by the vegetation itself was estimated at 50 metric tons, so that gross primary productivity was about 80 metric tons per hectare per year. Compared with temperate forests, these values are approximately 2.5 times higher for net productivity and 4 times higher for gross productivity, the difference being that the respiration rate at the tropical site was 5 times that of temperate forests.

Despite the overall high rates of productivity and biomass in tropical rainforests, the growth rates of their timber trees are not unusually fast; in fact, some temperate trees and many smaller herbaceous plants grow more rapidly. The high productivity of tropical rainforests instead results in their high biomass and year-round growth. They also have particularly high levels of consumption by herbivores, litter production, and especially plant respiration.

EATING THE RAINFOREST: HERBIVORY AND HOW PLANTS DEFEND THEMSELVES

Herbivory, the consumption of plant materials (generally leaves, shoots, and stems) by animals, is a defining process in most plant communities and a major influence on plant assemblages in tropical forests. Rainforest vegetation is under constant attack by hordes of sap drinkers, leaf eaters, leaf scrapers, leaf cutters, leaf miners, stem borers, shoot miners, and other types. More specifically, these herbivores include larvae and adults of the insect orders Lepidoptera (butterflies and moths), Hymenoptera (bees, wasps, and ants), and Coleoptera (beetles), including tortoise beetles, as well as adult or immature Heteroptera and Homoptera (the true bugs and other plant-sucking insects). Many insects, especially lepidopterans, are specialists, feeding only on a specific species, genus, or family of plants. On the other hand, orthopterans (grasshoppers, katydids, crickets, and roaches) can be more indiscriminate feeders. Mammalian herbivores include spiny rats, deer, peccaries, sloths, monkeys, and many others; they are often generalists, feeding on a variety of available plant taxa according to season or locality. Both insect and mammalian herbivores can influence tree demographics by the consumption of tree seedlings.

Herbivory is countered by plants through a myriad of defenses. Classical defenses include the production of defensive chemicals, such as alkaloids or aromatic terpenes, or other defensive substances, such as the entrapping latex produced by the breadnut and rubber trees native to South America. Defensive structures include toughened leaves, crystalline substances (oxalic acids) within plant tissues, trichomes (hairy projections), or spines and thorns. The trunks of *Astrocaryum* palms, for example, are densely covered with spines up to 30 cm (12 inches) long. Defensive coloration is a strategy used by some plants, the leaves of which always appear unhealthy because of their yellow shade. Defensive mutualisms include ant defense of cecropias against caterpillars and other insects. Plants also use a variety of more sophisticated defenses against herbivory, including the production of decoy butterfly eggs by some passion-flowers.

The majority (up to 70 percent) of leaf herbivory in the tropics occurs on young leaves, which are high in nitrogen and water and are relatively easy to eat because they are soft. For this reason, many plants exhibit higher levels of chemical defense in their developing tissues than in mature tissues, which are usually defended by structural means instead. In addition, most plants can be divided into two groups: those that yield many new leaves at once and thereby satiate herbivores through their synchronous flushing, or leaf production, and those that yield only a few new leaves at a time, carefully protecting these leaves with large allocations of chemical defense. In the first case, plants often "cheapen" the new leaves by delaying the allocation of metabolically "expensive" compounds such as chlorophyll until new leaves have toughened and are relatively protected. In many plants, fast growth comes at the expense of good defense; for example, plants that colonize canopy gaps first, such as balsa and cecropia, are often affected severely by insect herbivores.

The Status of the World's Tropical Forests

As recently as the 19th century tropical forests covered approximately 20 percent of the dry land area on Earth. By the end of the 20th century this figure had dropped to less than 7 percent. The factors contributing to deforestation are numerous, complex, and often international in scope. Mechanization in the form of chain saws, bulldozers, transportation, and wood processing has enabled far larger areas to be deforested than was previously possible. Burning is also a significant and dramatic method of deforestation. At the same time, more damage is being done to the land that is the foundation of tropical forest ecosystems: heavy equipment compacts the soil, making regrowth difficult; dams flood untouched tracts of wilderness to produce power; and mills use wood pulp and chips of many tree species, rather than a select few, to produce paper and other wood products consumed primarily by the world's industrialized nations. Although political, scientific, and management efforts are under way to determine means of slowing the destruction of tropical forests, the world's remaining acreage continues to shrink rapidly as demand for wood and land continues to rise.

The Global Implications of Deforestation

The implications of forest loss extend far beyond the borders of the states in which the forests grow. The role that rainforests play at the global level in weather, climatic change, oxygen production, and carbon cycling, while significant, is only just beginning to be appreciated. For instance, tropical rainforests play an important role in the exchange of gases between the biosphere and atmosphere. Significant amounts of nitrous oxide, carbon monoxide, and methane are released into the atmosphere from these forests. This metabolism is being changed by human activity. More than half the carbon monoxide derived from tropical forests comes from their clearing and burning, which are reducing the size of such forests around the world.

Another consequence of deforestation must be examined. In the upper Amazon River basin of South America, the rainforest recycles rains brought primarily by easterly trade winds. Indeed, surface transpiration and evaporation supply about half the rainfall for the entire region, and in basins of dense forest far from the ocean such local processes can account for most of the local rainfall. Should the Amazon Rainforest, which accounts for 30 percent of the land area in the equatorial belt, disappear, drought would likely follow, and the global energy balance might well be affected.

The Effects of Population Growth

The primary forces causing tropical deforestation and forest degradation can be tied to economic growth and globalization and to population growth. Population growth drives deforestation in several ways, but subsistence agriculture is the most direct in that the people clearing the land are the same people who make use of it. Rural populations must produce what food they can from the land around them, and in the rainforest this is most often accomplished via slash-and-burn agriculture. Forest is cleared, the cuttings are burned, and crops are planted for local consumption. However, the infertile tropical soils are productive for only a few years, and so it is soon necessary to repeat the process elsewhere. This form of shifting agriculture has been practiced sustainably among aboriginal cultures worldwide for centuries. Small patches of forest are cleared and abandoned when they become unproductive. The community then settles another isolated part of the forest, thus allowing previously settled land to regenerate.

However, in areas throughout the tropics larger populations than before now live at the forest margins. As subsistence agriculture progresses onto adjacent land, there is no opportunity for regeneration, especially if the shifting population is increasing. In some regions lowland forests have already been exhausted, and upland forests have been cleared. Land located on the slopes of hills and mountains is particularly susceptible to erosion and, therefore, to loss of the topsoil needed to sustain vegetation—arboreal or agricultural. Lowland tropical forests are not immune to erosion, however, as the heavy rainfall washes away unprotected soils.

Another subsistence-related factor in deforestation is demand for fuelwood, which is the main source of energy for 40 percent of the world's population. As population increases, this demand exerts significant and growing pressure on tropical forests, particularly in Africa.

Resettlement Programs

Urban population growth has led to the establishment of resettlement programs in several countries. Governments have made land available to poor families in overcrowded cities, who then have attempted to begin new lives from cleared forest. In Brazil the Transamazonian highway system was begun in the 1960s to enable development and settlement of the Amazon Rainforest. Part of the Transamazonian highway, called BR 364, penetrates the remote state of Rondônia in west-central Brazil. Since the highway's construction, this region has undergone significant deforestation. Main roads are cut into the forest, and parallel sets of access roads allow access to individual plots of land that are settled by farmers. This method of settlement results in a characteristic "fishbone" pattern when the land is viewed from above.

FORESTS

Small farms line the slopes in the highlands of Burundi, one of the most densely populated regions in Central Africa. Dr. Nigel Smith/The Hutchison Library

Brazil's resettlement program, while extensive, is by no means the largest. Population resettlement to provide agricultural employment and access to land is also important in some Southeast Asian countries, notably Indonesia, Malaysia, and Vietnam. By far the largest program has been conducted in Indonesia, where more than four million people have been voluntarily resettled from Java and Bali to the less-populated islands, especially to the province of Irian Jaya on the island of New Guinea. Despite considerable success, the program has been plagued by such problems as improper site selection, environmental deterioration, migrant adjustment, land conflicts, and inadequate financing. A program in Malaysia has been quite successful, in part because it set much smaller settlement targets and was better funded. Vietnamese development policy also utilized the resettlement of people in an effort to revitalize areas outside the major population centres.

While resettlement in Malaysia or Indonesia entails sea travel to isolated islands, roads connect South American population centres to the Amazon, where frontier cities draw both unsuccessful farmers from rural areas and migrants from established cities. The Amazon basin has long been relatively uninhabited, but improved diets and sanitation and the greater ease of transportation are making it more attractive for human settlement. From the mid-1940s onward, a number of "penetration roads" have been built from the populous highlands of Colombia, Ecuador, Peru, and Bolivia into Amazonia, often in conjunction with Brazil's Transamazonian highway. These roads have funneled untold numbers of landless peasants into the lowlands. Its vast area notwithstanding, the Amazon basin by the late 20th century had a predominantly urban population. Almost one-third of the estimated

Forests

nine million Brazilians living in the 4.9 million-square-km (1.9 million-square-mile) area officially designated as Legal Amazonia were concentrated in Belém and Manaus, each with more than one million inhabitants, and in Santarém. These cities, which are logistic bases of operations for cattle ranching, mining, timber, and agroforestry projects, are still growing rapidly, with modern residential towers and shantytowns standing side by side. Even frontier trading centres in the interior, such as Marabá, Pôrto Velho, and Rio Branco, have 100,000 or more inhabitants. In the upper reaches of the drainage area, places such as Florencia in Colombia, Iquitos and Pucallpa in Peru, and Santa Cruz in Bolivia have become significant urban centres.

This map shows Brazil and its extensive river systems. Encyclopædia Britannica, Inc.

Ranching and Mining

Most of those who come to the Amazon in resettlement programs are ill-prepared to become frontier farmers in an environment so naturally unsuitable to field agriculture, and the plots are soon abandoned. But the forest does not often reclaim the land; it is usually taken over by cattle ranchers first. In the Amazon and Central America the single largest use of cleared land is beef production—most of it for export. Cattle ranching thus illustrates how economic growth and globalization drive deforestation; other examples include logging and mining.

Tropical forests throughout the world often grow atop rich mineral deposits that are most easily mined by first clearing away the forest. The minerals are then extracted and sold in the global marketplace by the governmental or corporate enterprises involved. Even small tropical islands such as Fiji and New Caledonia have not been immune to deforestation by mining. In addition to clearing forests to gain access to deposits, mining also adds to deforestation by taking wood from the surrounding forest for ore processing. Such is the case in the Carajás region of Brazil, where tropical forest trees fuel iron smelters.

Gold deposits have been found in Indonesia and Papua New Guinea, as well as in the tropical forests north and south of the Amazon River. The resulting Amazon "gold rush" has brought as many as a half million transient miners (*garimpeireos*) equipped with picks, shovels, and sluice boxes to search for the mineral in alluvial deposits. Brazil's annual production peaked in 1987 at nearly 90 tons, declining thereafter. Meanwhile, the mercury used in extracting the gold polluted waterways, causing the fish that are so important in the local diet to become inedible. On the Madeira River teams operating from rafts pump auriferous sediments from the riverbed; the sediments are subjected to a similar treatment.

Short-Term Interests Versus Long-Term Gains

Ostensibly, countries possessing tropical forests seek sources of trade, such as mining and logging, and income to raise their populations' standard of living. It is often argued, however, that the underlying cause of economic dilemmas facing these governments is that control of resources is too concentrated among a wealthy few. Furthermore, these decision makers are not always from the developing countries, as multinational corporations can wield substantial influence on developing or unstable economies.

A common denominator in the destruction of tropical forests worldwide has been the pursuit of short-term

Large iron mine in the Serra dos Carajás, Pará state, Brazil. © Tony Morrison/South American Pictures

gains at the expense of long-term prospects, both economic and environmental. By the end of the 20th century the importance of tropical forests had been realized, and conservation had become a subject of international politics. The institutional arrangements controlling tropical forests began to change significantly as the roles of environmental and other nongovernmental organizations (NGOs) at local, national, and international levels expanded. Recent changes have resulted in some measure of progress: development projects have been halted; sustainable management programs have become a focus of research; developing countries have established governmental departments to oversee the use of natural resources; and a broader range of interest groups, such as indigenous tribal peoples, are being considered. Protected areas are being set aside throughout the world as cooperation between institutions at the international level is realized. In 1997, for example, Brazil established 57,000 square km (22,000 square miles) of land as protected rainforest in the state of Amazonas, creating the world's largest rainforest reserve.

Ecotourism

The recent emergence of the ecotourism industry is a phenomenon that relies on the cooperation of various groups with interests in tropical forests. Ecotourism is recreational travel for the purposes of observing and experiencing natural environments. Rainforests are popular destinations, and these sites are often jointly operated by a combination of governmental, private, environmental, and indigenous groups. Ecotourism facilities also serve as biological research stations, and vice versa. In this way ecotourism can be seen as contributing to conservation efforts.

Concerns for the Future

Such changes, while encouraging, are only beginning to work against the continuing decrease in acreage. International agreements among governments and businesses are highly dependent on the cooperation and commitment of the parties involved. Enforcement of policies at all levels of government, both within and between countries, is problematic. The record extent of fires in Amazonia and Indonesia in 1997–98 underscored profound problems in spite of recent progress. The relationships between oftentimes competing groups—local, national, and international; economic and environmental; governmental and nongovernmental—are what will determine the future of the planet's tropical forests.

THE TEMPERATE FOREST

A temperate forest is a vegetation type with a more or less continuous canopy of broad-leaved trees. Such forests occur between approximately 25° and 50° latitude in both hemispheres. Toward the polar regions they grade into boreal forests, which are dominated by evergreen conifers, so that mixed forests containing both deciduous and coniferous trees occupy intermediate areas. Temperate forests usually are classified into two main groups: deciduous and evergreen.

Deciduous forests are found in regions of the Northern Hemisphere that have moist, warm summers and frosty winters—primarily eastern North America, eastern Asia, and western Europe. In contrast, evergreen forests—excepting boreal forests, which are covered in boreal forest—typically grow in areas with mild, nearly frost-free winters. They fall into two subcategories: broad-leaved forests and sclerophyllous forests. (Sclerophyllous vegetation has small, hard, thick leaves.) The former grow in regions that have reliably high, year-round rainfall; the latter occur in areas with lower, more erratic rainfall. Broad-leaved forests dominate the natural vegetation of New Zealand; they are significantly represented in South America, eastern Australia, southern China, Korea, and Japan; and they occur in less well-developed form in small areas of southeastern North America and southern Africa. Sclerophyllous forests occur particularly in Australia and in the Mediterranean region.

The Origin of the Temperate Forest

Temperate forests originated during the period of cooling of world climate that began at the start of the Cenozoic Era (65.5 million years ago). As global climates cooled, climatic gradients steepened with increasing latitude, and areas with a hot, wet climate became restricted to equatorial regions. At temperate latitudes, climates became progressively cooler, drier, and more seasonal. Many plant lineages that were unable to adapt to new conditions became extinct, but others evolved in response to the climatic changes, eventually dominating the new temperate forests. In areas that differed least from the previously tropical environments— where temperate evergreen forests now grow—the greatest numbers of plant and animal species survived in forms most similar to those of their tropical ancestors. Where conditions remained relatively moist but temperatures dropped in winter, deciduous trees evolved from evergreen rainforest ancestors. In areas that became much more dry—though not to the extent that tree development was inhibited and only scrubland or desert environments were favoured— sclerophyllous trees evolved.

During the rapid climatic fluctuations of the past two million years in which conditions alternated between dry, cold glacial states—the ice ages of some northern temperate regions—and warmer, moister interglacial intervals, tree species of temperate forests had to migrate repeatedly to remain within climates suitable for their survival. Such migration was carried out by seed dispersal, and trees that were able to disperse their seeds the farthest had an advantage. In the North American and European regions where ice-sheet development during glacial intervals was most extensive, the distances that had to be traversed were greatest, and many species simply died out. Extinctions occurred not only where migration distances were great but also where mountains or seas provided barriers to dispersal, as in southern Europe. Thus, many trees that were formerly part of the European temperate forests have become extinct in the floristically impoverished forest regions of western Europe and are restricted to small refuge areas such as the Balkans and the Caucasus. For example, buckeye (*Aesculus*) and sweet gum (*Liquidambar*) are two trees that no longer occur naturally in most parts of Europe, having disappeared during the climatic turmoil of the past two million years.

Human activities have had pronounced effects on the nature and extent of modern temperate forests. As long ago as 8,000 years, most sclerophyllous forests of the Mediterranean region had been cut over for timber or cleared to make space for agricultural pursuits. By 4,000 years ago in China the same process led to the removal of most broad-leaved and deciduous forests. In Europe of 500 years ago the original deciduous forests had disappeared, although they are remembered in nursery tales and other folklore as the deep, wild woods in which children and princesses became lost and in which dwarfs and wild animals lived.

The deciduous forests of North America had been cleared almost completely by the end of the 19th century. Australia and New Zealand experienced similar deforestation about the same time, although the earlier activities of pre-European peoples had substantial impacts. The character of the Australian sclerophyllous forests changed in response to more than 38,000 years of burning by the Aboriginal people, and the range of these forests was expanded at the expense of broad-leaved forests. In New Zealand about half the forested area, which previously had covered almost the entire country, was destroyed by fire brought to the island by the Polynesian inhabitants who arrived 1,000 years before the Europeans.

Temperate Forest Environments

Winter in the temperate latitudes can present extremely stressful conditions that greatly affect the vegetation. The days are shorter and temperatures are low, so much so that in many places leaves are unable to function for long periods and are susceptible to damage from freezing. These conditions reduce the photosynthetic activity of the trees. In regions where winter temperatures regularly fall well below the freezing point and where soil moisture and nutrients are not in short supply, many trees have evolved a type of leaf that is relatively delicate and thin with a life span of a single growing season. Because such deciduous leaves do not require a large input of chemical energy, it is not too wasteful for the plant to shed them after a single growing season.

The "throwaway" leaves of temperate deciduous trees are shed as the days shorten in autumn and are replaced by new leaves in the spring. The forests dominated by these trees, therefore, have extremely pronounced seasonal changes in appearance, function, and climate. Most trees in temperate deciduous forests follow this habit, although some evergreen species usually are scattered among them. In particular, several broad-leaved evergreen shrubs are found in the understory of temperate deciduous forests that have less delicate, longer-lasting leaves than their deciduous neighbours; these leaves have adaptations that allow them to survive the freezing winter temperatures, and they can carry out photosynthesis for more than a single summer.

In areas where milder conditions prevail, however, photosynthesis may be possible at any season without need for protective mechanisms against frost damage. In these relatively unstressful circumstances most trees may benefit from retaining their leaves throughout the year, and heavy use of resources through frequent leaf replacement is thereby avoided. In environments such as these that also have a sufficient supply of moisture, temperate broad-leaved forests are found.

In mild but drier temperate environments, moisture shortage necessitates that trees develop thickened leaves. These leaves often have a reduced surface area or they dangle pendulously from limbs, two strategies employed to slow the loss of water (transpiration). High levels of energy and nutrients are needed to produce these thick leaves, which, therefore, cannot be replaced easily at annual intervals, an added reason for sclerophyllous trees to retain their leaves throughout the year. With foliage perennially present, these trees can carry out photosynthesis whenever moisture becomes available, provided temperatures are warm enough; this characteristic is advantageous where rainfall is infrequent and unpredictable.

Soils in temperate sclerophyllous forests are frequently poor in mineral nutrients. This poverty of nutrients accounts in part for the nature of the vegetation, because the annual production of leaves or the development of a dense broad-leaved canopy requires a significant input of nutrients. In contrast, soils in regions of deciduous and broad-leaved evergreen forest are generally fertile. (The forests that occupied the best soils in most regions, however, have been cleared almost completely to make way for agriculture.) Typical temperate deciduous forest soils are mull soils, which have a high level of organic matter especially close to the surface that is well mixed with mineral matter. Variations in soil materials and fertility have a strong influence on the types of trees that will dominate the forest. For example, in northwestern Europe, the European beech (*Fagus sylvatica*) dominates deciduous forests on shallow soils that overlie chalk, while oak (*Quercus*) is dominant on deeper, clay soils. The richness of the ground flora beneath the trees generally increases with soil fertility.

Intimate associations, or mycorrhizae, between tree roots and fungi are important and occur in most tree species. Although important in all forest types, these interactions have been studied more thoroughly in temperate deciduous forests. The fungal component of this symbiotic partnership grows on or in the fine roots of trees and benefits by obtaining nutrition in the form of carbohydrates from the tree root; the tree, in turn, is better nourished because a mycorrhizal root is more efficient at absorbing dissolved mineral nutrients from the soil than is an uninfected root.

Waterlogging of soils in temperate deciduous woodlands commonly occurs in regions with higher rainfall and humidity in late winter and spring, such as the British Isles. This occurs not only because in winter rainfall is higher and evaporation lower but also because the trees, barren of foliage, transpire a minimal amount of moisture. Areas subject to waterlogging include clay-rich soils, places that have low slope angles, depressions, and spots along watercourses. These places tend to have a richer ground flora but a less luxuriant tree canopy, which consists of only a few species that are tolerant of wet soils.

The depth of tree roots in temperate deciduous forests varies, but in many instances roots are shallow, with few reaching 1 metre (3.28 feet) below the surface. In the European beech, for example, shallow lateral growth of roots predominates over the development of a deep taproot, leading to growth of a "root plate" just beneath the soil surface. This enables the tree to exploit nutrients released at the surface by litter decomposition efficiently, while avoiding deeper layers that may become waterlogged. However, trees with root plates are more prone to being blown over in gales, especially after heavy rain has made the soil more plastic.

Bracket fungi, which grow on tree trunks, are among some of the largest fungi. Some species may reach 40 cm (16 inches) in diameter. H.S. Knighton

The Biota of Temperate Forests

The principal regions of deciduous forest all occur in the Northern Hemisphere and have historical connections between them. Thus, many similarities exist among their biota. The same plant and animal genera tend to occur in all regions, although different species are found in each region. However, the European deciduous forest flora is poorer than that of eastern North America and East Asia. Many plants are common and widespread in the forests of North America and Asia, but in Europe they are present only as restricted relict populations or fossils. Examples include hickory (*Carya*), *Magnolia*, sour gum (*Nyssa*), and sweet gum.

THE DECIDUOUS FOREST

Deciduous forests are composed primarily of broad-leaved trees that shed all their leaves during one season. Deciduous forest is found in three middle-latitude regions with a temperate climate characterized by a winter season and year-round precipitation: eastern North America, western Eurasia, and northeastern Asia. Deciduous forest also extends into more arid regions along stream banks and around bodies of water.

Oaks, beeches, birches, chestnuts, aspens, elms, maples, and basswoods (or lindens) are the dominant trees in mid-latitude deciduous forests. They vary in shape and height and form dense growths that admit relatively little light through the leafy canopy. Shrubs are found primarily near clearings and forest edges, where more light is available, and herbaceous flowering plants are abundant within the forest in the spring, before the trees come into full leaf.

The soils upon which deciduous forests thrive are gray-brown and brown podzols. They are slightly acidic and have a granular humus layer known as mull, which is a porous mixture of organic material and mineral soil. Mull humus harbours many bacteria and invertebrate animals such as earthworms.

Snails, slugs, insects, and spiders are common inhabitants of the deciduous forest, and many cold-blooded vertebrates, such as snakes,

Forests

Deciduous forest in fall coloration, Wasatch Mountains, Utah. Dorothea W. Woodruff—EB Inc.

frogs, salamanders, and turtles, are also present. Birds are represented by warblers, flycatchers, vireos, thrushes, woodpeckers, hawks, and owls. Prominent mammals include mice, moles, chipmunks, rabbits, weasels, foxes, bears, and deer.

Differences in temperature, moisture, and elevation may cause the formation of distinct plant associations within the deciduous-forest pattern. The dominance of beeches and maples in the northern part of the eastern North American deciduous forest and that of oaks and hickories along the southern extension of this vegetation are typical examples.

Flora

Most of the areas of North American deciduous forest are dominated by oaks (several species of *Quercus*) and/or beech (*Fagus grandifolia*), with maples (*Acer*) and species of hickory and linden or basswood (*Tilia*) also widespread. Beech and basswood are rare in other North American vegetation types, but oaks, hickories, and maples are more widespread.

In addition to these widespread species, many other trees are important components of the North American deciduous forests on a local scale. The rich, mixed mesophytic (adapted to environments that are neither too dry nor too wet) forest type found north of the Appalachians includes buckeye and tulip tree (*Liriodendron*), while the southern floodplain forest of the Mississippi valley is made up of oaks mixed with sour gum and the evergreen conifer swamp cypress (*Taxodium*). Southward and eastward of the Appalachians, oaks and hickories mingle with another conifer, pine (*Pinus*), while east of the Great Lakes, beech mix with maples, birch (*Betula*), and hemlock (*Tsuga*). Elsewhere, ash (*Fraxinus*), hop hornbeam (*Ostrya*), poplar (*Populus*), elm (*Ulmus*), and, until its decimation by fungal infection, chestnut (*Castanea*), are also important. A wide range of understory shrubs and small trees includes dogwood (*Cornus*), holly (*Ilex*), *Magnolia* species, and serviceberry (*Amelanchier*).

Fragments of the North American deciduous forest also occur on mountains in Mexico and Guatemala, where many of the same trees—such as *Fagus*, *Fraxinus*, *Juglans* (walnut), *Liquidambar*, *Quercus*—occur as identical or closely related species. Commonly these trees are accompanied by an understory of evergreen shrubs of tropical affinity.

In Japan, Korea, and China, north of the evergreen broad-leaved forests there is a gradual transition to deciduous forests. In Japan deciduous forests are dominated by beech (*Fagus crenata* and *F. japonica*), oak (*Quercus crispula*), and maple (*Acer carpinifolium* and other species); other trees mingle with these, including cherries (several species of *Prunus*), ash, *Magnolia*, and, in the east, the evergreen conifer fir (*Abies*). The leaves of many deciduous trees in Japan, like those in North America but unlike most in Europe, turn to bright shades of red and yellow before they are shed in autumn, the maples being particularly spectacular. Below the trees a dense layer of dwarf bamboo (*Sasa*) commonly grows; it may be so thick that it prevents the canopy trees from regenerating from seedlings. Thus, rapid, dense regrowth by dwarf bamboo may seriously interfere with reforestation after logging. Many small flowering herbs such as *Aconitum*, *Shortia*, *Mitchella*, and *Viola* grow at ground level. Much of the area of Chinese deciduous forests is dominated by various oaks, frequently mixed with other trees including maples, alder (*Alnus*), ash, walnut, poplar, and many others. A varied understory includes many small trees or shrubs such as hornbeam (*Carpinus*), dogwood, service-tree (*Sorbus*), and the shrubs *Acanthopanax* and *Aralia*, which are relatives of ivy. Deciduous forests of birch fringe the oak forests at their northern and montane margins.

The dominant trees of the European deciduous forests are generally closely related to their equivalents in North America and Asia, consisting of different species of common genera. In well-drained areas, such as those on sloping ground or permeable soils, most deciduous forests are dominated by European beech or by one of a few species of oak. Beech is overwhelmingly dominant across large areas that have an oceanic climate; these regions are damper and milder than most other regions in which deciduous forests grow because of the influence of prevailing winds from the Atlantic Ocean. Eastward from Britain and western France through central Europe into Russia there is a progressive decrease in this oceanic influence. Along this gradient the forest flora changes. For example, beech, durmast oak (*Quercus petraea*), and European hornbeam (*Carpinus betula*) all reach their eastward limits in the area north of the Carpathian Mountains, while linden (*Tilia cordata*) is typical of a small number of deciduous forest trees that extend east beyond Moscow.

Places subject to seasonal waterlogging in European deciduous forests—and which are also nutrient rich—often are dominated by alder (*Alnus glutinosa*), growing above a rich ground flora including ferns (*Athyrium* and *Dryopteris*), sedges (*Carex*), and forbs (*Caltha* and *Filipendula*).

The only significant temperate deciduous forests in the Southern Hemisphere occur in a small area of Chile around Valdivia, between about 36° and 41° S. Forests here are dominated by a deciduous species of beech, *Nothofagus obliqua*, which usually grows amid evergreen trees more typical of the broad-leaved forests bordering this area to the south.

The milder environments that support temperate evergreen forests generally lie closer to the Equator than do areas with temperate deciduous forest. They have richer biotas than the sclerophyllous or deciduous forests that grow in more stressful environments at similar latitudes, although they are less rich than the tropical rainforests where environmental stress is at a minimum throughout the year.

In the southernmost regions of Japan and Korea, which enjoy a warm, wet climate, the natural vegetation is evergreen broad-leaved forest dominated by oaks and their near relative *Castanopsis cuspidata* and by the laurel *Machilus thunbergii*. Camphor laurel (*Cinnamomum camphora*), figs (*Ficus retusa*), pandans (*Pandanus boninensis*), palms (*Livistona subglobosa*), and other plants that require year-round warmth also occur in the warmest places, whose vegetation is described as subtropical by some authorities and as warm temperate forest by others. Shrubs in these forests include species of *Aucuba*, *Camellia*, and *Eurya*, and orchids and ferns are commonly found at ground level.

Similar forests in southern China, now almost completely replaced by farmland or tree plantations, had an even richer flora, with oaks, members of the laurel family, and the tea relative *Schima* being prominent examples of a large diversity of trees. The remnants of these Chinese evergreen broad-leaved forests extend west to the foothills of the Himalayas, where similar forests at lower altitudes include many trees such as *Alcimandra*, *Castanopsis*, *Machilus*, *Magnolia*, and *Mallotus*. At higher altitudes there are more sclerophyllous forests that contain fewer species and are dominated by oaks.

In Australia a variety of temperate broad-leaved forests occur, usually as small patches in moist, sheltered, and fire-protected areas in mountainous and coastal terrain along the east coast. They are described as temperate rainforests owing to similarities with the flora, structure, and ecology of tropical rainforests that are found in similar environments to the immediate north. The warm temperate rainforests of milder, more northern areas display a high diversity of trees, including coachwood (*Ceratopetalum apetalum*), crab apple (*Schizomeria ovata*), and yellow carabeen (*Sloanea woollsii*). Palms are often present, as are various climbing plants and epiphytes (plants that grow on other plants but that derive moisture and nutrients from rain), although not to the extent that they occur in tropical rainforests. Ferns are typically abundant, and many large, graceful tree ferns grow there.

A similar warm temperate rainforest grows in northern parts of New Zealand; it contains a mixed broad-leaved canopy of trees such as *Elaeocarpus*, *Metrosideros*, and *Weinmannia*, which is frequently penetrated and overtopped by tall conifers, including the massive kauri (*Agathis australis*). Palms (*Rhopalostylis sapida*) and various lianas are often present.

Cool temperate evergreen broad-leaved forests in the southernmost areas of eastern Australia, particularly Tasmania, and in New Zealand and the southern portion of South America are usually dominated by evergreen species of beech (*Nothofagus*), with different species occurring in each region. Few other trees typically coexist with *Nothofagus* in these cool forests, which also lack climbers and vascular epiphytes, although they may have a great abundance of mosses on tree trunks, branches, and sometimes leaves. In Australia other trees that may be present include sassafras (*Atherosperma moschatum*), *Elaeocarpus holopetalus*, and leatherwood (*Eucryphia lucida*), while in New Zealand conifers in the plum pine family (Podocarpaceae) commonly emerge above the broad-leaved canopy, especially on sites with a history of natural disturbance such as landslides. In Chile other trees growing with *Nothofagus*, or in some cases forming temperate evergreen forests without it, include *Eucryphia*, Chilean laurel (*Laurelia*), and *Persea*, with bamboos (*Chusquea*) becoming abundant at some sites after forest disturbance.

South of the European deciduous forests lie areas that were occupied by temperate sclerophyllous forests before the effects of human manipulation of the environment were felt. These areas extend as a narrow ring around the coastline of the Mediterranean Sea. Typical evergreen trees are oaks (several species including the cork oak, *Quercus suber*) and the pistachio relative *Pistacia lentiscus*, commonly mixed with various deciduous species near their northern limits and with pines elsewhere. Forests similar to the Mediterranean sclerophyllous forests extend east, discontinuously, as far as the western slopes of the Himalayas, although throughout their range they have been extensively altered by human activities and nowhere are considered to be in their natural state.

Temperate sclerophyllous forests are most widespread in Australia. It is remarkable that across the major portion of a very large area, the most common trees are various species of one genus, *Eucalyptus*. About 500 species are known of this archetypal Australian tree, which occurs naturally in few other regions. However, gum trees, as they are often called, are well known in many regions to which they have been introduced. Two other diverse and widespread trees in these forests are *Acacia* and *Casuarina*. Many other trees, shrubs, and grasses occupy subordinate positions. A sharp boundary separates the temperate sclerophyllous forests from the tropical and temperate rainforests of moister sites in the east. This abrupt transition is maintained by fire, which prevents or reverses invasion of the sclerophyllous forest by rainforest plants. However, where fires are infrequent, especially in southern regions, rainforest vegetation can form an understory beneath the *Eucalyptus* canopy. To the north, temperate sclerophyllous forests grade into similar tropical savanna and sclerophyllous forests; inland they merge into shrublands and deserts.

Fauna

The fauna of temperate forests resembles the regional fauna. However, the structure of the vegetation provides similar ecological niches in all regions of the same vegetation type, so that, although different species inhabit different forests, they are of a similar type. Tree holes provide homes and nest sites for arboreal mammals and birds in most regions of temperate forest but with pronounced variations. For example, apart from bats no native mammals are found in the New Zealand forests. In Australia the arboreal mammals are all marsupials or bats, including gliders such as the greater glider (*Petaurus volans*) and opossums such as the common ringtail (*Pseudocheirus peregrinus*), which nests in holes, and the well-known koala (*Phascolarctos cinerea*), which is free-living and feeds mainly or entirely on young tree foliage.

In temperate forests of the Northern Hemisphere, squirrels are widespread. Local additional arboreal forms in Asian forests include monkeys, most of which are predominantly seedeaters. This feeding niche is particularly appropriate in Northern Hemisphere forests, which include more trees with large seeds, such as the acorn-producing oaks, than do their Southern Hemisphere equivalents.

Birds are less regionally distinct, with families such as those of the owl and pigeon being well represented in almost all temperate forest regions. Nevertheless, there are still some pronounced regional variations. The tits (*Paridae*) dominate the foliage-gleaning insectivore guild in Europe, where warblers (*Sylviidae*) are less varied; this situation is reversed in North America. More fundamental contrasts are apparent in Australia, where honeyeaters, which feed on nectar, and parrots, which feed on small, hard seeds, are diverse and common in the sclerophyllous forests. In the Northern Hemisphere few plants provide nectar for birds, and tree seeds are usually eaten by squirrels and pigeons.

THE THORN FOREST

Thorn forests are made up of dense, scrublike vegetation characteristic of dry subtropical and warm temperate areas with a seasonal rainfall averaging 250 to 500 mm (about 10 to 20 inches). This vegetation covers a large part of southwestern North America and southwestern Africa and smaller areas in Africa, South America, and Australia. In South America, thorn forest is sometimes called *caatinga*. Thorn forest grades into savanna woodland as the rainfall increases and into desert as the climate becomes drier.

A thorn forest consists primarily of small, thorny trees that shed their leaves seasonally. Cacti, thick-stemmed plants, thorny brush, and arid-adapted grasses make up the ground layer. Many annual plants grow, flower, and die during the brief rainy season.

Population and Community Development and Structure of Temperate Forests

At mid-latitudes the sun never rises to the near-vertical position in the sky as it does in the tropics. In winter, when the sun appears particularly close to the horizon throughout the short period of daylight, its direct radiant energy impinges most on slopes facing the Equator—southward in the Northern Hemisphere, northward in the Southern Hemisphere. In deciduous forest regions, primarily in the Northern Hemisphere where the trees are without leaves in winter, direct sunlight bathes the forest floor. The ground surface and the vegetation on southern slopes are quickly warmed, paradoxically creating a more stressful environment for plants than exists on the shadier, cooler northern slopes. The stress is due to more extreme temperature fluctuations than the southern slope sustains, which affects the speed at which the plant tissues, frozen during the cold night, are thawed. On the southern slope the thawing is much more rapid, which is more damaging than the gradual thaw that the vegetation on the northern slopes experiences. As a result, herbaceous plants of the forest floor that retain green leaves throughout winter, such as *Hepatica*, *Hydrophyllum*, and *Tiarella* in eastern North America, are more common on northern slopes, while sunnier slopes become bare and brown during the winter months.

As days lengthen and temperatures warm in spring, new, green photosynthetic shoots develop rapidly from buds that formed and food reserves that were laid down in storage tissues during the previous growing season. Annuals growing from seed are rare. The most rapid development of all typically occurs in the herbaceous plants of the forest floor. These plants must take immediate advantage of the spring warmth and sunlight before the new tree foliage casts its heavy shade over the ground, drastically reducing available light energy and slowing the rate of photosynthesis. Many plants of the forest floor have underground energy-storing organs such as bulbs, corms, or fleshy rhizomes that allow them to grow rapidly and strongly in spring to gain maximum advantage from the short, warm, shade-free period. They frequently also produce strong, pointed shoots that are able to emerge above the thick layer of dead leaves that dropped from the trees of the canopy the previous autumn. Thus, the ground of temperate deciduous forests in spring typically is covered by a green carpet of foliage that often includes dramatic displays of colourful flowers before the buds on the trees above have opened. Later, after the leaves of the tree canopy have regrown, the ground cover declines. Typical low-growing plants in North American deciduous forests include species of *Cypripedium*, *Erythronium*, *Hydrophyllum*, *Trillium*, and *Viola*. In Europe bluebells (*Hyacinthoides non-scripta*), daffodils (*Narcissus pseudonarcissus*), and wood anemones (*Anemone nemorosa*), among many other herbaceous species, provide similarly spectacular spring flower displays.

In autumn the delicate leaves of the deciduous trees senesce and start to die. As this happens, they lose their green colour and turn various shades of brown, yellow, and red. Dramatic displays are created, becoming significant tourist attractions in the areas in which the colours are brightest—that is, eastern North America and western Asia. A weak layer of tissue called the abscission layer develops at the base of each leaf stalk, and at this point the stalk breaks and the leaf is shed. The massive leaf drop that ensues during autumn has earned the season its alternate designation, fall.

Forests

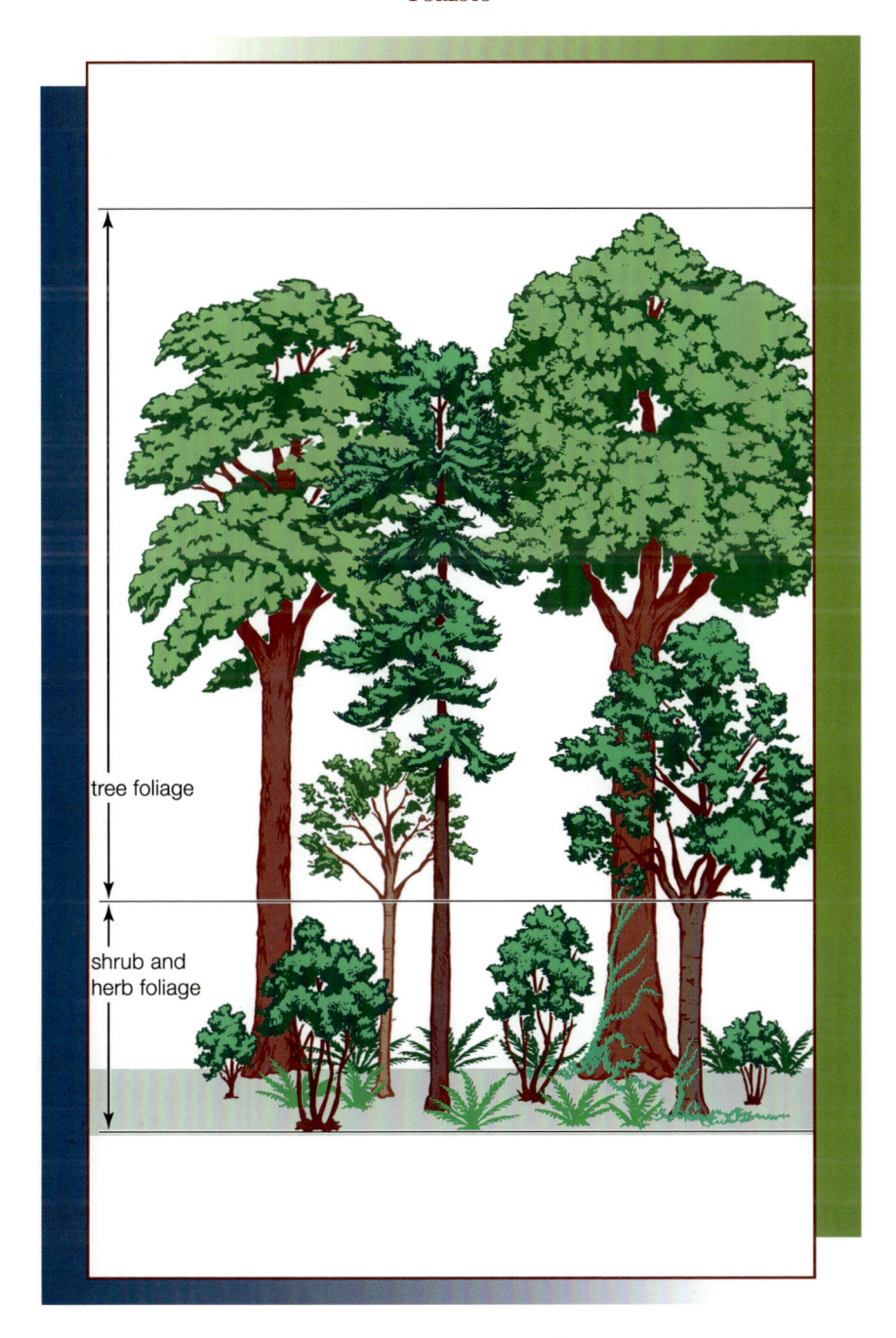

Vegetation profile of a temperate deciduous forest. Encyclopædia Britannica, Inc.

As in other forests, the composition of the temperate deciduous forest is commonly determined by the influence of disturbances—natural as well as human—on tree regeneration. For instance, in eastern North America, the tulip tree (*Liriodendron tulipifera*) produces seeds that can remain dormant in the soil for up to seven years. When they germinate, subsequent establishment of the quick-growing saplings is most successful on bare mineral soil in full light. Stands of this species therefore tend to be of the same age and act as markers of the time at which a catastrophe destroyed the former tree cover, baring the ground. Beech that occur in the same region are shade tolerant and occupy places not subject to catastrophe, regenerating beneath undamaged tree canopy.

There are still a few temperate forests that have been disturbed little by human activities, and the interrelationships between the vegetation and the large animals are of interest. A rare, relatively intact area of deciduous forest that contains some evergreen conifers is found in Poland. The most common trees include linden (species of *Tilia*), oak (*Quercus robur*), hornbeam (*Carpinus betulus*), maple (*Acer platanoides*), and spruce (*Picea abies*). Until 1923 large areas were dominated by linden, but most tree regeneration at that time was by spruce, hornbeam, and maple. By 1973 these latter species had become common as canopy trees, and lindens began regenerating freely, as they had not at the earlier time. Such changes result in part from the varying regeneration requirements of the trees. Most successful regeneration takes place where gaps of light reach the forest floor. Competition between saplings of different species is intense and has different outcomes depending in part on soil and light intensities and to a large degree on the effects of mammals. Rooting by wild pigs, although it destroys many small plants, creates suitable conditions for seedling establishment. Saplings are subject to browsing by deer, which also feed on herbaceous plants that compete with tree seedlings. European bison similarly eat foliage and, in addition, can debark trees with their horns. The combined variable influences of these and other animals interact to produce a mosaic forest of different tree ages and composition, with longer-term changes resulting from population fluctuations due, in part, to human impact.

Most temperate forests, where they still survive at all, have been so exploited and disturbed by human influences that their natural condition is difficult to discern. However, some areas outside the old agricultural regions of Asia and Europe have sustained much less human disturbance, making it possible to define natural disturbances responsible for some variations. For example, temperate broad-leaved forests in Chile have been shown to vary in structure and composition according to their history of natural catastrophe in the form of earthquake-induced landslides; similar relationships between ecosystem alteration and natural disturbances have been demonstrated in New Zealand with volcanic eruption and in Tasmania with wildfire. In all these regions, sites undisturbed for many centuries have forests dominated by shady, highly competitive species of *Nothofagus*, often with few seedlings of any kind beneath the large, old trees. However, in the wake of natural catastrophe, other trees can invade the sites, and only gradually does *Nothofagus* reestablish itself and slowly resume dominance during subsequent tree generations. Therefore, in areas that have suffered many instances of disturbance, there exists today a variety of forest types. Comparable variations resulting from storm damage and wildfire occurrence have been recognized in North American deciduous forests.

The Biological Productivity of Temperate Forests

The total aboveground biomass (dry weight of organic matter in an area) for temperate deciduous forests is typically 150 to 300 metric tons per hectare; values for temperate broad-leaved forests are generally higher, and those for sclerophyllous forests are lower. The subterranean component is more difficult to measure, but it appears to approximate a value of about 25 percent of the aboveground component in deciduous forests and rather more in temperate evergreen forests. Total biomass in temperate deciduous forests is, therefore, about 190 to 380 metric tons per hectare. These values refer to mature, undamaged forests; when disturbances also are considered, the range would include lower values.

For temperate forests gross primary productivity (the total biomass fixed by the vegetation in a unit area within a unit time) has been estimated at 16 to 50 metric tons per hectare per year. Net primary productivity, gross primary productivity less that used by plants in respiration, is approximately 10 metric tons per hectare per year; it is greatest in young forests where the trees are rapidly growing toward full size, and it declines in forests of old trees. While the gross primary productivity of temperate forests is considerably lower than that of tropical rainforests, the net primary productivity is not so different, reflecting the lower diversity and complexity of the consumer component of the temperate forest ecosystem.

Temperate forests have been useful to human populations in many diverse ways. Although in most places they have been replaced by simpler agricultural systems, large areas still remain, especially on poorer soils, and are important sources of timber. The total yield as well as the quality of timber is maximized by keeping forests in a condition of greatest net primary productivity—that is, by harvesting trees before they reach their age of declining growth. Old growth forests have a high biomass and considerable conservational significance but are not efficient in terms of total sustained timber yield.

THE BOREAL FOREST

Boreal forests, or taiga, are a vegetation type composed primarily of cone-bearing, needle-leaved, or scale-leaved evergreen trees, found in regions that have long winters and moderate to high annual precipitation.

The boreal (meaning northern) forest region occupies about 17 percent of the Earth's land surface area in a circumpolar belt of the far Northern Hemisphere. Northward beyond this limit, the boreal forest merges into the circumpolar tundra. The boreal forest is characterized predominantly by a limited number of conifer species—that is, pine (*Pinus*), spruce (*Picea*), larch (*Larix*), fir (*Abies*)—and to a lesser degree by some deciduous genera such as birch (*Betula*) and poplar (*Populus*). These trees reach the highest latitudes of any trees on the Earth. Boreal plants and animals are adapted to short growing seasons of long days that vary from cool to warm. Winters are long and very cold, the days are short, and a persistent snowpack is the norm. The boreal forests of North America and Eurasia display a number of similarities, even sharing some plant and animal species. The northern forests of Russia, especially Siberia, are referred to as taiga, meaning "little sticks," a term now widely accepted as an alternative to boreal forest.

The Origin of the Boreal Forest

During the final period of maximum cold temperatures (23,000 to 16,500 years ago), in the latter part of the

Pleistocene Ice Age (which ended 11,700 years ago), species that now constitute the boreal forest were displaced as far south as 30° N latitude by the continental glaciers of Europe, Asia, and North America and by the hyperarid and extremely cold environments of unglaciated Asia and North America. As the glaciers began to retreat gradually about 18,000 years ago, species of the boreal forest began to move northward in Europe and North America. In eastern and central North America the northward movement of the forest was relatively steady and gradual. An exception to this progression occurred about 9,000 years ago in western Canada when white spruce spread rapidly northward across 2,000 km (1,240 miles) of newly deglaciated land in only 1,000 years. This rapid migration resulted from seed dispersal facilitated by strong northward winds caused by clockwise atmospheric circulation around the remnant ice cap of northern Quebec and the western part of Hudson Bay.

Because so much of the Earth's water was bound up in ice at this time, sea levels were lower than they are today, and this allowed migrations of various terrestrial species to occur. Many areas that are now islands were then connected to the nearby mainland—for example, the British Isles were linked to Europe. As the climate warmed during the last stages of the glacial period, but before the sea level rose to its current position, some plants and animals of the mainland European boreal forest migrated to Britain. This biota exists today as part of the boreal forest in the highlands of Scotland. The areas of lowland central Alaska, the central Yukon territory, and far eastern Russia, which had climates too arid to permit the formation of ice sheets, were connected by the Bering Land Bridge, across which many species migrated. As a result, today across Alaska a gradient in plant characteristics can be observed, ranging from typical North American forms in the east to those with Eurasian characteristics in the west.

The Distribution of the Boreal Forest

The boreal forests of North America and Eurasia are broad belts of vegetation that span their respective continents from Atlantic to Pacific coasts. In North America the boreal forest occupies much of Canada and Alaska. Although related transition forest types are present in the northern tier of the lower 48 United States, true boreal forest stops just north of the southern Canadian border. The vast taiga of Asia extends across Russia and southward into northeastern China and Mongolia. In Europe most of Finland, Sweden, and Norway are covered with boreal forest. A small, isolated area of boreal forest in the Scottish Highlands lacks some continental species but does contain the most widespread conifer of the Eurasian boreal forest, Scotch pine (*Pinus sylvestris*).

The position of the boreal forest zone generally is controlled by the degree of warmth experienced during the growing season, the temperature of the soil, and the extreme minimum winter temperature. The boreal forest belt consists of three roughly parallel zones: closed canopy forest, lichen woodland or sparse taiga, and forest-tundra. The closed canopy forest is the southernmost portion of the taiga. It contains the greatest richness of species, the warmest soils, the highest productivity, and the longest growing season within the boreal zone. North of the closed canopy forest is the lichen woodland—a smaller parallel zone of sparse forest or woodland in which tree crowns do not form a closed canopy. Lichen mats and tundralike vegetation make up a significant portion of the ground cover. To the north of the lichen woodland lies forest-tundra, which occurs along the northern edge of tree growth (tree line). Patches of trees consisting of only a few species dot restricted portions of the landscape, forming a complex mosaic with tundra. Many trees in the forest-tundra zone have never been known to produce viable seeds or have done so only sporadically. These trees were established during warmer climatic episodes from a few hundred to a few thousand years ago and have persisted since, usually by vegetative (asexual) reproduction. Forest fires in this zone remove trees, and because of the lack of reproduction, only unburned patches of trees remain.

The closed forest, or southern taiga zone, on both continents is not distributed along a strictly east-west axis. At the western margin of Europe the warming influence of the Gulf Stream allows the closed canopy forest to grow at its northernmost location, generally between about 60° and 70° N. In western North America the Kuroshio and North Pacific currents likewise warm the climate and cause the northward deflection of the forest into Alaska and Yukon in Canada. On the eastern margin of the continents the boreal forest is deflected southward to between about 50° and 60° N by the cold polar air masses that flow south along these coasts. This is the southernmost limit of the boreal forest, to the south of which, in humid eastern North America and Europe, lies a northern deciduous broad-leaved transition forest. In this forest small stands of boreal conifers are distributed on cooler or less productive sites such as peaty wetlands. In the arid centre of both continents the closed canopy boreal forest is bordered to the south by a forest parkland of trees and grassland.

The central portions of Eurasia and North America are regions of flat or gently rolling topography. There, the northern and southern boundaries of the boreal forest are broad and gradual; they have fluctuated by as much as 200 km (125 miles) during the past few thousand years. A well-defined but complex boundary is formed between taiga and alpine tundra on the mountains of the Pacific edge in western North America and the far eastern region of Russia. Generally the boreal forest does not come into contact with the humid temperate or subpolar rainforest of coastal Alaska and British Columbia because of high mountain barriers, but some low-elevation regions have a transition zone often characterized by trees that are a hybrid of Sitka spruce (*Picea sitchensis*) and white spruce (*P. glauca*). In Norway and Scotland a variant form of the boreal forest occupies extremely humid environments.

Practically all the large river systems of the taiga of Siberia, including the Ob, Yenisey, and Lena rivers, are northward flowing. The Ob in western Siberia forms a great lowland basin with a considerable percentage of the land surface covered with poorly drained peaty wetlands. In such situations within the boreal zone a closed canopy forest is generally absent.

The Climate of the Boreal Forest

Coldness is the dominant climatic factor in boreal forest regions, although a surprising diversity of climates exists. Several factors—namely, the solar elevation angle, day length, and snow cover—conspire to produce this cold climate. In the boreal region the Sun is never directly overhead (90°) as it can be in the tropics. The maximum solar angle decreases with increasing latitude. At latitude 50° N in the southern boreal region the maximum solar angle is 63.5° and at the Arctic Circle it is only 47°. As a result solar energy is less intense in the boreal regions because it is spread out over a greater area of the Earth's surface than it is in equatorial regions. Day length also affects temperature. Long winter nights at high latitudes allow radiation emitted by the surface of the Earth to escape into the atmosphere, especially in continental interiors where cloud cover is less abundant than it is near the coast. Snow cover, too, affects the climate, because it reflects incoming solar radiation and amplifies cooling.

During winter, a snowpack persists for at least five months in the southern portion of boreal forest regions and for seven or eight months in the northern reaches. The boreal forest actually mitigates this cooling because it roughens and darkens what would otherwise be a smooth, snow-covered, energy-reflecting surface for much of the year. It has been estimated that the Earth would be significantly colder without the boreal forest.

The northern limit of the North American boreal forest coincides with the mean position of the Arctic front—the boundary between Arctic and mid-continental air masses—in the summer; its southern limit coincides with the mean frontal position in the winter. Mean annual temperatures in the boreal forest range from a few degrees Celsius above freezing to -10 °C (14 °F) or more. Areas with a mean annual temperature below freezing are susceptible to the formation of permafrost soils (frozen ground).

The mean temperature of January, the coldest month, is generally less than -10 °C (14 °F) across the boreal region. The boreal forest includes areas that experience some of the lowest temperatures on the Earth, excluding Antarctica. At the height of winter an intensely cold pocket of air develops over inland areas of far eastern Siberia; mean temperatures of -50 °C (-58 °F) have been recorded in this region. As this Siberian cold air flows over the unfrozen northern Pacific Ocean a great temperature contrast develops that results in strong, westward-moving storm systems. The movement, position, and strength of these storms control much of the weather in the Northern Hemisphere.

Boreal forests do not grow on areas surrounding the Bering Strait. A rigorous cold climate with a very short snow-free season precludes the growth of trees on the Russian side of the Bering Strait in the Chukotka region of the Russian Far East. On the North American side in western Alaska summers are too cool for trees to grow because of cold air masses moving off the Bering Sea.

The growing season in the boreal forest is generally cool; the mean temperature of the warmest month, July, is generally between 15 and 20 °C (59 and 68 °F). Summer daytime high temperatures are typically cool to warm—20 to 25 °C (68 to 77 °F)—for much of the growing season in the boreal forest. Cool summer temperatures can actually produce higher photosynthetic efficiency in plants than can warmer conditions. Species adapted to cold respire less (use up less of their food stores) when photosynthesizing at cool temperatures in intense summer light than they do at higher temperatures, allowing a greater net gain in biomass (dry mass of organic matter).

Areas of the boreal forest located in the centre of continents generally receive 30 to 50 cm (12 to 20 inches) of

THE CONIFEROUS FOREST

Coniferous forests constitute a vegetation type composed primarily of cone-bearing, needle-leaved, or scale-leaved evergreen trees, found in regions of the world that have long winters and moderate to high annual precipitation. The northern Eurasian coniferous forest is called the taiga, or the boreal forest. Both terms are used to describe the entire circumpolar coniferous forest with its many lakes, bogs, and rivers. Coniferous forests also cover mountains in many parts of the world. Pines, spruces, firs, and larches are the dominant trees in coniferous forests. They are similar in shape and height and often form a nearly uniform stand with a layer of low shrubs or herbs beneath. Mosses, liverworts, and lichens cover the forest floor.

The light-coloured, usually acidic soils of coniferous forests are called podzols (podsols) and have a compacted humus layer, known as mor, which contains many fungi. These soils are low in mineral content, organic material, and number of invertebrate animals such as earthworms.

Coniferous forests and lakes on the ancient Baltic Shield of Finland. H. Fristedt/Carl E. Ostman ab

Mosquitoes, flies, and other insects are common inhabitants of the coniferous forest, but few cold-blooded vertebrates, such as snakes and frogs, are present because of the low temperatures. Birds include woodpeckers, crossbills, warblers, kinglets, nuthatches, waxwings, grouse, hawks, and owls. Prominent mammals include shrews, voles, squirrels, martens, moose, reindeer, and wolves.

Eurasian coniferous forest is dominated in the east by Siberian pine, Siberian fir, and Siberian and Dahurian larches. Scots pine and Norway spruce are the important species in western Europe.

North American coniferous forest is dominated throughout by white spruce, black spruce, and balsam fir, although lodgepole pine and alpine fir are important species in the western section.

A distinct subtype of the North American coniferous forest is the moist temperate coniferous forest, or coast forest, which is found along the west coast of North America eastward to the Rocky Mountains. This subtype is sometimes called temperate rain forest, although this term is properly applied only to broad-leaved evergreen forests of the Southern Hemisphere. Warm temperatures, high humidity, and often misty conditions encourage the development of a mossy, moisture-loving plant layer under the giant trees of Sitka spruce, western red cedar, western hemlock, Douglas fir, and coast redwood.

Other subtypes of coniferous forest occur at various elevations in the Rocky Mountains of North America, in Central America, and in eastern Asia. They are known as subalpine and montane forests and are dominated by combinations of pine, spruce, and fir species.

annual precipitation. Precipitation totals are relatively modest in these locations because they are a significant distance from unfrozen oceans that supply moisture. Some boreal forests are semiarid and may even include grasslands interspersed with the forest. These forests are found in regions of Yukon and Alaska that occur on the leeward side of mountains, which are sheltered from moisture-bearing winds, as well as in some portions of the interior of far eastern Russia. Annual precipitation in low elevations of these regions is 30 cm (12 inches) or less. The highest annual precipitation total in the boreal forest, which can exceed 100 cm (39 inches), is in eastern North America and northern Europe. During ancient eras of colder climate these regions also received relatively abundant precipitation, which resulted in the buildup of glacial ice sheets. Today these once heavily glaciated regions support extensive lakes, streams, and wetlands.

Extended periods of clear, dry weather in the boreal region are caused by persistent strong polar high pressure systems. If strong high pressure persists during the long days near the summer solstice, temperatures can warm to 30 °C (86 °F) or higher. Intense heating at the ground surface often produces convective storms with lightning but little rain, causing forest fires.

Boreal Forest Soils

Boreal forest conifer litter is highly acidic. Soils of the more humid and southern boreal forest are highly leached spodosols, which are characterized by the leaching of iron, aluminum, and organic matter from the chemically and biologically distinct surface layer—horizon A—to the next layer—horizon B. Much of the soil of central and eastern Canada—granitic Canadian Shield—has been repeatedly scraped clean by glacial advances. Thus, productive forests often are restricted to portions of the landscape where soil material has been deposited by glaciers. Peaty wetlands occur where surface drainage is impeded by permafrost, youthful glacial topography, or aggraded rivers; their soils are characteristically organic soils, or histosols. Soils in much of boreal western North America and Asia are inceptisols, which have little horizon development. Very thin surface salt deposits are found in the most arid portions of the boreal forest.

Cold soils are characteristic of the boreal forest region, which overlaps the zone of permafrost. Permafrost is soil or earth material that remains below 0 °C (32 °F) for at least two years. The surface, or active, layer of permafrost thaws in the warm season and freezes in the winter, but the soil below the active layer remains continuously frozen. Because the plant rooting zone is restricted to the active layer, nutrient supply is limited and secure anchoring for roots is lacking. Some trees and other plants of the taiga (especially black spruce [*Picea mariana*] and tamarack [*Larix laricina*] in North America and larches in Siberia) can grow on permafrost if the active layer is sufficiently deep, but several species are eliminated from permafrost.

The boreal forest itself is an important contributing factor to the development of permafrost. The latter stages of forest growth—characterized by development of an intact forest canopy, growth of an insulating moss cover in summer, and accumulation of forest litter—may cool the soil to such an extent that permafrost develops. Warming of the soil is promoted by forest fires, which remove the canopy, moss, and forest litter layers. In the absence of an intact canopy, a deeper and more effective insulating layer of snow accumulates in the winter. The presence of dark ash following a fire increases solar energy absorption on the site for several years.

The boreal forest of Europe generally lacks permafrost, but east of the Ural Mountains and from central Canada northward permafrost is common. In southern and central boreal forests, permafrost occurs sporadically and occupies only a small percentage of the landscape that experiences the coldest temperatures. The northern portion of closed canopy forest and the lichen woodland zone are in a region of discontinuous permafrost, where permafrost is found on north-facing slopes and in cold air drainage basins but is absent from south-facing slopes and newly deposited alluvial sites. Most of the forest-tundra is within the continuous permafrost zone.

Forest productivity in the middle and northern taiga zones is directly related to soil temperature. Warmer soils decompose organic matter more quickly, releasing nutrients for new plant growth and creating a more productive site. Productive forest types occupy warmer, south-facing slopes and river terraces, and less productive dwarf or sparse forest occupies the north-facing and basin permafrost sites.

Floodplains throughout the boreal forest regions are free of permafrost, high in soil fertility, and repeatedly disturbed in ways that renew the early, rapid growth stages of forest succession. Floodplains are a mosaic of productive shrubland and forest that serve as a major habitat for moose (*Alces alces*), which influence ecosystem structure and function.

South-central Alaska and adjacent Yukon and British Columbia support the most extensive ice sheets and glaciers in the world outside the polar desert regions of Antarctica and Greenland. Glacial meltwater is a large part of the flow of larger rivers such as the Yukon and Tanana in Alaska and the Yukon territory. Glacial meltwater carries a heavy load of suspended sediment that deposits in riverbeds and causes frequent channel shifts. Glacial river floodplains are extensive, very dynamic, and constantly renewed with fertile soil material. In the ancient past exposed deposits of glacial silt were picked up by strong winds and deposited on surrounding hillsides. Fertile soils, known as loess, resulted, on which highly productive upland forests are found today. Because the beds of glacially fed rivers are rising, the landscape through which they flow is partially drowned from the impeded drainage, often preventing forest growth and favouring the development of marshes and mires.

The Biota and Its Adaptations

Nearly all major taxonomic groups have fewer species of animals and plants in the boreal forest than they have in other terrestrial ecosystems at lower latitudes. This accords with the species diversity gradient that is observed from lower to higher latitudes, with numbers of species decreasing in a poleward direction.

Trees

Scotch pine is the most widely distributed pine species in the world, growing from northern Scotland to the Russian Pacific shore. The relatively humid and productive taiga of northern Europe and south-central Siberia is dominated by this species. Forest management has greatly favoured this species in Scandinavia and Finland. It is a thick-barked species and easily survives light ground fires, often reaching ages of 350 to 400 years, with some individuals being older than 700 years. European aspen and Siberian spruce are essentially transcontinental in distribution as well.

The species composition of Eurasian taiga is different east of central Siberia from that which prevails westward into Europe. Distinctive European species include Norway spruce (*Picea abies*), a large dominant species of the productive humid boreal forest, and Sukaczev larch (*Larix sukaczewii*), an early successional species (one of the first species to colonize an area after a disturbance) of European Russia. Gray (*Betula populifolia*) and white birch (*B. pendula*) occur across northern Europe and well into

Vegetation profile of a boreal forest. The tree layer consists mainly of conifers, and mosses are the predominant ground cover. Encyclopædia Britannica, Inc.

central Siberia. The birches often form dense stands of light- or white-barked trees that are considered a characteristic feature of the boreal forest. Siberian larch (*Larix sibirica*) and Siberian fir (*Abies sibirica*) are restricted to north-central Asia. Species restricted to northeastern Asia include chosenia (*Chosenia arbutifolia*), an early successional broad-leaved tree of floodplains; Siberian stone pine (*Pinus sibirica*), a short shrub or tree; and Asian spruce (*Picea obovata*).

All North American tree species are distributed across the continent except jack pine (*Pinus banksiana*), lodgepole pine (*Pinus contorta*), and balsam fir (*Abies balsamea*). Jack pine is a relatively small, short-lived, early successional tree occurring in the eastern and central parts of boreal forests east of the Rocky Mountains. Lodgepole pine is a longer-lived, early successional species growing in western Canada and along the Rocky Mountain axis from central Yukon southward to well south of the boreal forest limit. Balsam fir is a shade-tolerant, late successional, but relatively short-lived tree that occurs only in the eastern and central boreal forest.

Major taiga tree species are well adapted to extreme winter cold. The northernmost trees in North America are white spruce that grow along the Mackenzie River delta in Canada, near the shore of the Arctic Ocean. The northernmost trees in the world are Gmelin larch (*Larix gmelinii*) found at latitude 72°40′N on the Taymyr Peninsula in the central Arctic region of Russia.

Other Plants

A distinctive feature of the flora of boreal forests is the abundance and diversity of mosses; about one-third of the ground cover under boreal forest is dominated by moss. Much of the ground cover in older conifer stands is moss, which grows on rocks, on tree trunks, and in the pits formed by upturned trees. Extensive peaty wetlands in the boreal region are often thick accumulations of dead sphagnum and other mosses, sedges, and other plants; a living moss layer continually grows at the surface.

Lichens (a symbiotic association of a fungus and algae) constitute a significant part of the ground cover in the lichen woodland or sparse taiga. Lichens are also generally well distributed on tree trunks and especially in the canopy of older conifers throughout the boreal forest. Because lichens and mosses are dispersed by airborne spores that can travel long distances, many species of both groups are found across the entire circumpolar boreal forest.

Many vascular plants are also widespread across the circumpolar north. Some forest understory species dominate their habitats; they include twinflower (*Linnaea borealis*), lingonberry (*Vaccinium vitis-idaea*), baneberry (*Actaea rubra*), and Swedish and Canadian dwarf cornel (*Cornus suecica* and *C. canadensis*). Several boreal forest plants are adapted to rapid colonization and growth in recently burned areas, such as fireweed (*Epilobium angustifolium*). The extensive peatlands of the boreal north support a typical flora that usually includes species such as Labrador tea (*Ledum palustre*), cloudberry (*Rubus chamaemorus*), cotton grass (*Eriophorum* species), and crowberry (*Empetrum nigrum* or *E. hermaphroditum*). In northern Europe crowberry also grows as shrub mats under Scotch pine forests or woodlands. Crowberry has been shown to produce secondary chemical compounds that inhibit or kill Scotch pine seedlings. Periodic light ground fires reduce the abundance and vigour of crowberry and allow tree regeneration.

Specialized orchids in the forest understory include calypso (*Calypso bulbosa*), coral root (*Corallorrhiza trifida*), and lady's slipper (*Cypripedum* species). The roots of these plants form particular associations with fungi (mycorrhizae). Willow shrubs (*Salix* species) are one of the first plants to emerge following disturbances on floodplains and occasionally on uplands as well. Important grasses across the boreal region include species of bromegrass (*Bromus* species), bluegrass (*Poa* species), reed bent grass (*Calamagrostis* species), and vanilla grass (*Hierochloe odorata*). Many freshwater aquatic plants such as sedges (*Carex* species) and pondweeds (*Potamogeton* species) are distributed widely across the boreal zone of both continents because migratory waterfowl and shorebirds are effective in dispersing their seeds. Several species of ferns are common to the boreal forests of the two continents, especially in regions of higher precipitation.

Mammals

Because a winter snowpack is a dependable feature of the taiga, several mammals display obvious adaptations to it. The snowshoe, or varying, hare (*Lepus americanus*), for example, undergoes an annual change in colour of its pelage, or fur, from brownish or grayish in the summer to pure white in the winter, providing effective camouflage. Its feet are large in proportion to its body size, a snowshoelike adaptation for weight distribution that allows the hare to travel over the surface of snow rather than sink down into it. The lynx (*Lynx canadensis*) is the principal predator of the snowshoe hare. It, too, has large feet, with fur between the toes, enabling the lynx to remain on the snow's surface. Most animals of the boreal forest are well adapted to the cold and survive it easily if they have enough food to maintain an energy balance through the winter.

Moose are the largest browsing animals in the boreal forest. In the summer they eat willow and broad-leaved trees and also wade in lakes and ponds to consume aquatic plants. Throughout the winter moose eat large quantities of woody twigs and buds. Moose depend on high-quality feeding areas in the shrub zone along river floodplains and on the early successional growth of woody plants in burned or cutover forest. Intensive browsing by moose can alter the composition of the forest in its early stages of growth, often increasing the dominance of conifers, which they do not consume in as great amounts as they do broad-leaved trees. Harvesting a moose for winter food is an important and even critical element of survival for humans living in isolated rural areas of the taiga.

Moose populations are controlled by various means. Wolves (*Canis lupus*) prey on moose across most of the boreal forest, and some scientists and game managers

Population Fluctuations: Lynx-Hare Cycle

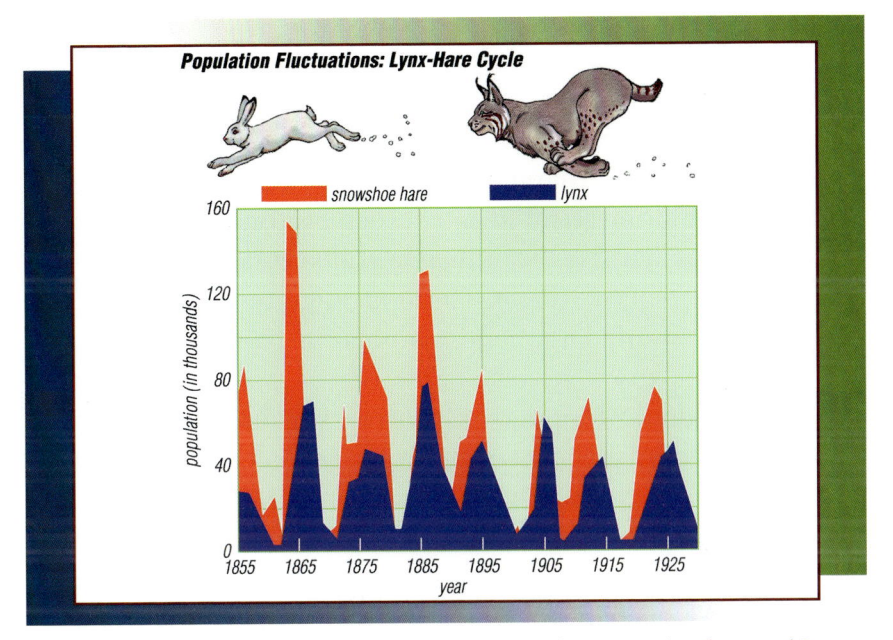

Cyclical fluctuations in the population density of the snowshoe hare and its effect on the population of its predator, the lynx. The graph is based on data derived from the records of the Hudson's Bay Company. Encyclopædia Britannica, Inc.

believe that once moose numbers are depressed wolf predation can keep moose populations low. As a result, wolf trapping or shooting programs are carried out as a game management measure to increase prey numbers. The natural regulation of moose populations by wolf predation and the presence of wolves themselves is valued as well. As a result programs to control wolf populations are often the subject of intense debate. Other factors control moose numbers, such as the restriction of access to plants during years of deep snow and lack of early successional woody plant growth caused by forest maturation. Where the boreal forest is extensively cut for forest products, moose numbers have increased greatly, often to levels that are considered undesirable for forest regeneration. Subsistence and sport hunting of moose are widely used tools of moose population management.

Another large-hoofed browsing mammal that is present seasonally in the boreal forest is the reindeer (*Rangifer tarandus*) in Eurasia and the closely related caribou in North America. A large portion of the reindeer population is semidomesticated and herded by nomadic peoples such as the Sami of Scandinavia and several native peoples in northern Russia. Caribou migrate the greatest distances of any large land mammal in North America. They often move in vast herds of 500,000 animals or more, seldom stopping or pausing because they must constantly forage in these environments of generally low productivity. During the early winter, reindeer and caribou migrate south from their summer ranges in the tundra to the forest-tundra or lichen woodland, where they graze primarily on lichens. Later in winter caribou typically move to open forests and sedge-rich lake margins or to snow-free wind-swept mountains. In April and

American mink (Mustela vison). Karl H. Maslowski

May, caribou form long columns and migrate back north to the tundra.

Several mammals of the boreal region are valued for their furs, and trapping and trade in furs has been an important part of the culture, economy, and history of the region as long as humans have lived there. Important furbearing species include lynx and marten (*Martes americana*) and in wetland habitats beaver (*Castor canadensis*), mink (*Mustela vison*), and muskrat (*Ondatra zibethica*).

In the North American boreal forest the northern flying squirrel (*Glaucomys sabrinus*) is adapted to consume fungi, especially underground fruiting bodies (sporocarps) of fungi that form mutually beneficial relationships (mutualism) with trees by colonizing their roots. The flying squirrel's consumption and dispersal of these underground fungi provide a significant benefit to the forest as a whole.

BIRDS

The boreal forest is the migratory destination of large numbers of birds for the summer breeding season. These include several passerine songbirds typical of shrub and forest habitats, such as thrushes, flycatchers, and warblers. Many of these species consume insects in the canopy of the boreal forest and other habitats. Predators of these birds occur in the forest as well, such as the sharp-shinned hawk (*Accipiter striatus*) and the northern goshawk (*A. gentilis*). Populations of several boreal forest-breeding migratory thrushes, flycatchers, and warblers may be declining because of the loss of their wintering habitats in the tropical forests of the world and the changes to or loss of forest habitats in the temperate zones along their migratory routes.

Birds of the boreal forest fill a variety of niches. Some are seed consumers or dispersers, others are insect consumers. They carry out other specialized roles as well. For example, the yellow-bellied sapsucker (*Sphyrapicus varius*) drills evenly spaced rows of small holes in the bark of trees and then visits these "wells" to obtain sap and the insects it attracts. Various other birds, mammals, and insects benefit from the sap wells, too.

Woodpeckers excavate tree cavities, which subsequently are used by many species of birds and mammals. Woodpeckers are specialized predators of wood- and bark-inhabiting insects; they are thought to be important in the control of the spruce beetle (*Dendroctonus rufipennis*) population. In searching for insects, woodpeckers chisel or strip the bark off dead or dying trees, promoting more rapid decay and the release of nutrients from dead trees. As large old trees have become rarer through forest cutting, some year-round resident woodpeckers such as the northern three-toed woodpecker (*Picoides tridactylus*) and the great spotted woodpecker (*Dendrocopos major*) have lost their habitats and declined in numbers.

Few bird species remain in boreal forests through the long cold winters because of limited opportunities for food, although some undertake only a short migration south. Resident bird species include the common raven (*Corvus corax*) and the boreal and black-capped chickadees of North America and the Siberian tit (*Parus* species).

The extensive areas of lakes, ponds, and wetlands—especially in the glaciated part of the boreal forest—provide a large habitat for waterfowl and shorebirds, although the birds primarily occur in low densities across the landscape. North American shorebirds that breed in forested peatlands include common snipe (*Gallinago gallinago*) and yellowlegs (*Tringa* species). Commonly encountered waterfowl are northern pintail (*Anas acuta*), scaup (*Aythya* species), and scoters (*Melanitta* species).

Forests

Downy woodpecker (Dendrocopos pubescens). Kenneth and Brenda Formanek

Insects

The boreal forest is the home of relatively few species of insects, but extensive and usually uniform areas of habitat periodically support high populations of species that do live there. The boreal forest lacks the elaborate complexes of invertebrate predators and parasites that serve as stabilizers of the insect populations in warmer regions. As a result, boreal insect populations occasionally increase rapidly and cause outbreaks. Some outbreaks can injure or kill trees across widespread areas of the boreal forest. Once an outbreak reaches a certain size it can become self-sustaining, much like a forest fire; the effects of the spruce budworm and spruce beetle in North America are well-documented examples. Outbreaks can be triggered by unusual weather or physical injuries that stress trees and make them vulnerable to the insects; they can end for a variety of reasons, including production of defensive chemicals by the host plants or depletion of susceptible host plants.

Perhaps the insects most noticeable to humans in the boreal forest are mosquitoes, which belong to several species. Mosquitoes feed on and are fed upon by many of the birds of the boreal forest. Wetland areas of the boreal region, such as sites having poor drainage because of permafrost, provide extensive mosquito breeding sites. Where well-oxygenated, flowing water is found, biting flies are abundant. Almost all food webs that support fish in boreal forest streams are dependent on insects.

Conifers serve as hosts for a variety of wood-boring beetles, spruce beetles, bark beetles, and ips beetles (*Ips* species). These insects aid in wood decomposition and nutrient release. Some beetles have outer shells with specialized indentations specifically matched to the shape and size of the spores of wood-decomposing fungi. Fungal spores become securely lodged in these cuplike structures. As the beetles burrow into wood they inoculate it with fungi.

A variety of lepidopterans (moths and butterflies) are adapted to feeding on the leaves of boreal trees. These include defoliators and leaf rollers.

Soil Organisms

The species richness and total biomass of soil organisms are significantly lower in the boreal forest than they are at lower latitudes. Dominant soil organisms are protozoans, nematodes, rotifers, and tardigrades. These organisms live primarily in soil water film and soil pore water. The soil fauna of the boreal forest region is distinctive because it generally lacks large invertebrates such as millipedes, isopods (springtails), and earthworms, especially in the middle and northern taiga. Larger soil invertebrate animals perform the function of biting off (shredding) pieces of leaf litter in forest soils and passing them through their guts. As a result of this activity thick layer of several years' accumulation of only partially decomposed plant material is characteristic of boreal forest soils.

Fungi are the dominant organisms in the task of decomposition of boreal forest litter, but flushes of bacterial growth occur in response to triggering factors. The soil animals generally do not attack the forest litter directly but instead exert their influence by grazing on the fungi and bacteria. The rate of decomposition in boreal forest soils does not keep pace with the rate of production, causing the progressive accumulation of organic matter. At middle depths of the forest floor small invertebrates, especially dipteran larvae, partially consume or skeletonize leaf litter before emerging as adults.

The Community Structure of Boreal Forests

The structure of the biological community in boreal forests is determined by the severity and frequency of natural and human-caused disturbances. Forest fires and logging are the largest sources of ecological disturbance in this biome.

Natural Disturbances

The boreal forest is well adapted to development following natural disturbances, which include fire, floods, snow breakage, and insect outbreaks. Characteristic of the boreal forest is the general lack of late successional species that develop under an intact forest canopy.

Fire is the primary agent responsible for natural disturbances in the boreal forest. It can result from natural causes, such as lightning, or it can be set by humans. Large-scale insect outbreaks can weaken or kill trees over vast areas, thus creating an environment less resistant to fire. In the period between 1981 and 1989 an estimated 3 million hectares (7.4 million acres) burned annually in the former Soviet Union, almost all of which occurred within the taiga region of Russia. The so-called Black Dragon Fire of 1987 in China and Russia may have been the largest single fire in the world in the past several hundred years. During the 20th century, about 1 million hectares (2.5 million acres) of taiga in Canada have burned annually; a great majority of the burning occurred in the less accessible boreal forest of the northern and western parts of the country. In Alaska in years that have prolonged hot and dry periods of summer weather, millions of hectares burn, primarily in a few very large fires. Intervals of about 200 years occur between fires in the uplands of northwestern Canada and in the interior of Alaska. In much of the central and western boreal forest of North America replacement of vegetation on upland sites, presumably by fire, appears to be necessary for forest regeneration. Floodplain islands usually do not burn and contain white spruce trees as old as 400 years. In the northern boreal forest of Europe a pattern of periodic light ground fires in Scotch pine forests was typical before the era of fire control. The thick bark of these mature trees allowed them to survive these fires. In much of the boreal forest only wildland fires that threaten high-value resources are actively suppressed. Complete fire suppression would cause soil temperature to decline gradually, promoting permafrost development that would cause a significant decrease in site productivity.

Jack pine and lodgepole pine have cones that remain closed on the tree (serotinous), and black spruce has semiserotinous cones; these cones do not open to release their seeds until a wax layer is melted by the heat of fire. White spruce seedlings require the bare mineral soil produced by burning of thick organic layers of the forest floor for proper establishment; they may time their periodic production of seed to dry periods when fire is more likely.

The Effects of Human Use and Management

Different degrees of forest development have had various effects on biodiversity around the circumpolar boreal forest zone.

A highly developed forest industry based on intensive forest utilization is maintained in boreal Scandinavian countries and Finland. About 95 percent of the productive forest types of Finland and the Scandinavian countries have been harvested at least once. Finland is located almost entirely within the boreal region and is one of the most forested nations in the world. About 6.5 percent of Finnish land, which includes large areas of marginal forest, woodland, and tundra, is protected from human modification. Only about 3 percent of Swedish forests are protected, most of which is concentrated in marginal forests of mountainous regions. Between 10 and 15 percent of species in Swedish forests are threatened.

The Canadian boreal forest represents nearly 7.5 percent of the Earth's forested area. Much of the harvesting of Canadian forest has been carried out in primary (previously unlogged) forest. Nearly all the mature first-growth timber, especially of southern, central, and eastern boreal Canada, is anticipated to be removed by the late 1990s. Considerable effort has been devoted to forest regeneration and tending of new stands, although a certain amount of land does not meet reforestation goals.

In Alaska the amount of land with at least 10 percent forest cover in the boreal region is estimated at about 46 million hectares (114 million acres), or 12 percent of the state, only 5.5 million hectares (13.6 million acres) of which is considered productive timberland. Of all areas in the world, Alaska probably has the largest percentage of its surface area, about 40 percent, devoted to strict protection of natural habitats and species. Local-scale logging traditionally has been carried out for much of the 20th century. Plans to accelerate logging are being considered.

The taiga of Siberia covers 680 million hectares (1.7 billion acres) and represents nearly 19 percent of the world's forested area and possibly 25 percent of the world's forest volume. About 400,000 hectares (990,000 acres) of the Russian taiga are logged annually, and nearly an equal area is burned, with perhaps half of the burned area resulting from destructive fires of human origin. Social and economic problems in the early postcommunist era have slowed the amount of logging by one-third to one-half; however, several large-scale joint ventures between Russian organizations and foreign partners to harvest the forests are under way. The fate of the Siberian boreal forest has become a matter of international concern.

Large areas, perhaps exceeding 2 million hectares (4.9 million acres), of the Russian taiga near Norilsk and the Kola Peninsula have been destroyed by air pollution. Many oil pipelines are leaking in Siberia, and repairs and maintenance are minimal. In July through September 1994 more than 150,000 metric tons of crude oil were spilled in the Kolva, Usa, and Pechora river basins of the republic of Komi in Russia. Spring meltwater could carry spilled oil in this region into the northward-flowing rivers that empty into the Arctic Ocean.

The Biological Productivity of Boreal Forests

Primary productivity (the rate at which photosynthesis occurs) of boreal forest ecosystems often is limited by cold soil temperatures. Net annual primary production (the total amount of productivity less that used by photosynthetic organisms in cellular respiration) in boreal forest types varies greatly, from slightly more than 2 metric tons per hectare near the polar tree limit to about 10 metric tons per hectare along its southern margin. Boreal forests are estimated to contain about 18 percent of the Earth's total biomass (the dry weight of organic matter). The boreal forest or taiga of Siberia alone represents 57 percent of the Earth's coniferous wood volume. Ecosystems and soils of the boreal region store a significant amount of the Earth's carbon in the form of dead but undecomposed or partially decomposed organic matter. Global warming or land use changes could enhance decomposition, leading to the release of increased amounts of stored carbon into the atmosphere in the form of the greenhouse gas carbon dioxide.

Deforestation

Although most of the areas cleared for crops and grazing represent permanent and continuing deforestation, deforestation can be transient. About half of eastern North America lay deforested in the 1870s, almost all of it having been deforested at least once since European colonization in the early 1600s. Since the 1870s the region's forest cover has increased, though most of the trees are relatively young. Few places exist in eastern North America that retain stands of uncut old-growth forests. In addition, while some forests are being cleared, some are being planted. The United Nations Food and Agriculture Organization (FAO) estimates that there are approximately 1.3 million square km (500,000 square miles) of such plantations on Earth. These are often of eucalyptus or fast-growing pines—and almost always of species that are not native to the places where they are planted.

Elsewhere, forests are shrinking. The FAO estimates that the annual rate of deforestation is about 1.3 million square km (500,000 square miles) per decade. About half of that is primary forest—forest that has not been cut previously (or at least recently). The greatest deforestation is occurring in the tropics, where a wide variety of forests exists. They range from rainforests that are hot and wet year-round to forests that are merely humid and moist, to those in which trees in varying proportions lose their leaves in the dry season, and to dry open woodlands.

Because boundaries between these categories are inevitably arbitrary, estimates differ in how much deforestation has occurred in the tropics.

Dry forests in general are easier to deforest and occupy than moist forests and so are particularly targeted by human actions. Worldwide, humid forests once covered an area of about 18 million square km (7 million square miles). Of this, about 10 million square km (3.9 million square miles) remained in the early 21st century. Given the current annual rates of deforestation, most of these forests will be cleared within the century. Indeed, in some places, such as West Africa and the coastal humid forests of Brazil, very little forest remains today.

The human activities that contribute to tropical deforestation include commercial logging and land clearing for cattle ranches and plantations of rubber trees, oil palms, and other economically valuable trees. Another major contributor is the practice of slash-and-burn agriculture, or swidden agriculture. Small-scale farmers clear forests by burning them and then grow their crops in the soils fertilized by the ashes. Typically, the land produces for only a few years and then must be abandoned and new patches of forest burned.

The Amazon Rainforest is the largest remaining block of humid tropical forest, and about two-thirds of it is in Brazil. (The rest lies along that country's borders to the west and to the north.) Detailed studies of Amazon deforestation from 1988 to 2005 show that the rate of forest clearing has varied from a low of about 11,000 square km (4,200 square miles) per year in 1991 to a high of about 30,000 square km (11,600 square miles) per year in 1995. The high figure immediately followed an El Niño, a repeatedly occurring global weather anomaly that causes the Amazon basin to receive relatively little rain and so makes its forests unusually susceptible to fires. Studies in the Amazon also reveal that 10,000 to 15,000 square km (3,900 to 5,800 square miles) are partially logged each year, a rate roughly equal to the low end of the forest clearing estimates cited above. In addition, each year fires burn an area about half as large as the areas that are cleared. Even when the forest is not entirely cleared, what remains is often a patchwork of forests and fields or, in the event of more intensive deforestation, "islands" of forest surrounded by a "sea" of deforested areas.

The effects of forest clearing, selective logging, and fires interact. Selective logging increases the flammability of the forest because it converts a closed, wetter forest into a more open, drier one. This leaves the forest vulnerable to the accidental movement of fires from cleared adjacent agricultural lands and to the killing effects of natural droughts. As fires, logging, and droughts continue, the forest can become progressively more open until all the trees are lost.

Although forests may regrow after being cleared and then abandoned, this is not always the case. About 400,000 square km (154,000 square miles) of tropical deforested land exists in the form of steep mountain hillsides. The combination of steep slopes, high rainfall, and the lack of tree roots to bind the soil can lead to disastrous landslides that destroy fields, homes, and human lives. Steep slopes aside, only about one-fourth of the humid forests that have been cleared are exploited as croplands. The rest are abandoned or used for grazing land that often can support only low densities of animals, because the soils underlying much of this land are extremely poor in nutrients. (To clear forests, the vegetation that contains most of the nutrients is often burned, and the nutrients literally "go up in smoke" or are washed away in the next rain.)

Deforestation has important global consequences. Forests sequester carbon in the form of wood and other biomass as the trees grow, taking up carbon dioxide from the atmosphere. When forests are burned, their carbon is returned to the atmosphere as carbon dioxide, a greenhouse gas that has the potential to alter global climate, and the trees are no longer present to sequester more carbon. In addition, most of the planet's valuable biodiversity is within forests, particularly tropical ones. Moist tropical forests such as the Amazon have the greatest concentrations of animal and plant species of any terrestrial ecosystem. Perhaps two-thirds of Earth's species live only in these forests. As deforestation proceeds, it has the potential to cause the extinction of increasing numbers of these species.

The coastal forest of Rio de Janeiro state, Braz., badly fragmented as portions were cleared for cattle grazing. Courtesy, Stuart L. Pimm

NOTABLE FORESTS OF THE WORLD

The world's forests range from the Equator to the high latitudes. Several individual forests are notable for the role they play in maintaining the global ecological balance, the economic and recreational resources they provide, or their value as settings of significant events in human history.

The Amazon and the Ituri

The Amazon and Ituri forests are textbook examples of tropical rainforests. The Amazon Rainforest is arguably the largest rainforest in the world, covering a vast portion of the northern part of South America. The Ituri, on the other hand, encompasses only a small part of Africa's vast Congo River basin. Despite their differences in size, both regions possess tremendous biological diversity.

The Amazon Rainforest

The Amazon Rainforest is a large, tropical rainforest occupying the drainage basin of the Amazon River and its tributaries in northern South America, and covering an area of 6 million square km (2.3 million square miles). Comprising about 40 percent of Brazil's total area, it is bounded by the Guiana Highlands to the north, the Andes Mountain Ranges to the west, the Brazilian central plateau to the south, and the Atlantic Ocean to the east.

Amazonia is the largest river basin in the world, and its forest stretches from the Atlantic Ocean in the east to the tree line of the Andes in the west. The forest widens from a 320-km (200-mile) front along the Atlantic to a belt 1,900 km (1,200 miles) wide where the lowlands meet the Andean foothills. The immense extent and great continuity of this rainforest is a reflection of the high rainfall, high

Canoe on the Negro River in the Amazon Rainforest, Amazonas state, northern Brazil. Union Press/Bruce Coleman, Inc., New York

humidity, and monotonously high temperatures that prevail in the region.

The Amazon Rainforest is the world's richest and most varied biological reservoir, containing several million species of insects, plants, birds, and other forms of life, many still unrecorded by science. The luxuriant vegetation encompasses a wide variety of trees, including many species of myrtle, laurel, palm, and acacia, as well as rosewood, Brazil nut, and rubber tree. Excellent timber is furnished by the mahogany and the Amazonian cedar. Major wildlife includes jaguar, manatee, tapir, red deer, capybara and many other types of rodents, and several types of monkeys.

In the 20th century, Brazil's rapidly growing population settled major areas of the Amazon Rainforest. The Amazon forest shrank dramatically as a result of settlers' clearance of the land to obtain lumber and to create grazing pastures and farmland. In the 1990s the Brazilian government and various international bodies began efforts to protect parts of the forest from human encroachment, exploitation, and destruction.

Plant Life

The overwhelmingly dominant feature of the Amazon basin is the tropical rainforest, or selva, which has a bewildering complexity and prodigious variety of trees. Indeed, as many as 100 arboreal species have been counted on a single acre of forest, with few of them occurring more than once. The Amazon forest has a strikingly layered structure. The sun-loving giants of the uppermost reaches, the canopy, soar as high as 40 metres (120 feet) above the ground; occasional individual trees, known as emergents, rise beyond the canopy, frequently attaining heights of 60 metres (200 feet). Their straight, whitish trunks are splotched with lichens and fungi. A characteristic of these giant trees is their buttresses, the basal enlargements of their trunks, which help stabilize the top-heavy trees during infrequent heavy winds. Further characteristics of the canopy trees are their narrow, downward-pointing "drip-tip" leaves, which easily shed water, and their cauliflory (the production of flowers directly from the trunks rather than from the branches). Flowers are inconspicuous. Among the prominent members of the canopy species, which capture most of the sunlight and conduct most of the photosynthesis, are rubber trees (*Hevea brasiliensis*), silk-cotton trees (*Ceiba pentandra*), Brazil nut trees (*Bertholletia excelsa*), sapucaia trees (*Lecythis*), and sucupira trees (*Bowdichia*). Below the canopy are two or three levels of shade-tolerant trees, including certain species of palms—of the genera *Mauritia*, *Orbignya*, and *Euterpe*. Myrtles, laurels, bignonias, figs, Spanish cedars, mahogany, and rosewoods are also common. They support a myriad of epiphytes (plants living on other plants)—such as orchids, bromeliads, and cacti—as well as ferns and mosses. The entire system is laced together by a bewildering network of woody ropelike vines known as lianas.

In addition to the rainforests of the *terra firme*, there are two types of inundated rainforests, *várzea* and *igapó*, which constitute about 3 percent of the total Amazon Rainforest. *Várzea* forests can be found in the silt- and nutrient-rich floodplains of whitewater rivers such as the Madeira and the Amazon, with their ever-changing mosaic of lakes, marshes, sandbars, abandoned channels, and natural levees. They are generally not as high, diverse, or old as those of the *terra firme*, and they are subject to periodic destruction by floods and human manipulation. (The *várzea* and its flood-free margins attract the most human settlement.) Wild cane (*Gynerium*) and aquatic herbs and grasses, as well as fast-growing pioneer tree species of the genera *Cecropia*, *Ficus*, and *Erythrina*, are conspicuous.

Igapó forests grow along the sandy floodplains of blackwater and clearwater rivers such as the Negro, the Tapajós, and the Trombetas. These forests may reach a maximum flood level of 12 metres (40 feet) for up to half the year, but they can be accessed by canoe.

The lowland rainforest on the Andean fringe grades into a discontinuous, tangled montane or cloud forest of misshapen trees cloaked with mosses, lichens, and bromeliads. There the cinchona, or fever-bark tree, once exploited for its antimalarial agent quinine, can be found. At still higher elevations is the grass and shrub growth of the *páramo* zones and cold Antiplano region.

Along the drier, southern margin of the Amazon basin, high forest gives way to the immense *cerrado* (scrub savanna), *campo* (grassy savanna), and *caatinga* (heath forest). The latter is characteristic of parts of the Mato Grosso Plateau, where taller forest is restricted to the stream courses and swales (marshy depressions) that dissect the upland surface. On the sandy soils of the lower Negro and the Branco drainage areas, and locally in Amapá, grassy savannas dotted with stunted trees replace the high forest. Large areas of grassy savannas can be found on the Mato Grosso Plateau, Marajó Island, the Llanos, and elsewhere.

Animal Life

To give a succinct overview of the complete fauna of Amazonia is as impossible as it is to adequately describe the great diversity of its flora; in part this is because many of the region's species have yet to be identified. The rivers and streams of the basin teem with life, and the forest canopy resonates with the cries of birds and monkeys and the whine of insects. There is a notable paucity of large terrestrial mammal species; indeed, many of the mammals are arboreal.

More than 8,000 species of insects alone have been collected and classified. Myriads of mosquitoes may transmit diseases including malaria and yellow fever. Leaf-cutter ants (of the genera *Atta* and *Acromyrmex*) are prevalent, as are the ubiquitous small black flies known as *piums* in Brazil. Fireflies, stinging bees, hornets, wasps, beetles, cockroaches, cicadas, centipedes, scorpions, ticks, red bugs, and giant spiders are abundant. Most spectacular, however, are the hundreds of species of brilliantly coloured butterflies; sometimes thousands of butterflies gather in the afternoon on riverside sandbanks.

About 1,500 fish species have been found within the Amazon system, but many more remain unidentified. Most fish are migratory, moving in great schools at spawning time. Fish represent a critical source of protein in the often meat-poor diet of the *caboclo* population. (The term *caboclo* refers to people of mixed European and Indian ancestry in Brazil who live off the rivers and forests.) Among the more important commercial species are the pirarucu (*Arapaima gigas*), one of the world's largest freshwater fish, and various giant catfish. The small, flesh-eating piranha generally feeds on other fish but may attack any animal or human that enters the water; its razor-sharp teeth cut out chunks of flesh, stripping a carcass of its meat in a few minutes. Some fish species have become locally threatened as the worldwide demand for frozen and dried fish has increased. Tropical fish of the Amazon are also some of the most desired for food and breeding and for use as aquarium specimens. Iquitos, Manaus, and the Colombian port of Leticia are centres of these trades.

Alligators are hunted for their skins; river turtles and their eggs are considered a delicacy; and the giant sea cow, or manatee, is sought for its flesh and for oil. All are threatened by overhunting, and the manatee has been listed as an endangered species. Aquatic animals also include river dolphins (*Inia geoffrensis*); the semiaquatic capybara, the largest rodent in the world (weighing up to 80 kg [170 pounds]); and the nutria, or coypu, valued especially for its pelt. Other common rodents are the paca, agouti, porcupine, and local species of squirrels, rats, and mice.

The tapir, the white-lipped peccary, and several species of deer are native to the Amazon basin and much sought for their meat. Water buffalo, introduced from Southeast Asia as work and dairy cattle, predominate in the remote, swampy parts of Marajó Island.

Especially characteristic of the Amazon forest are several species of monkeys. Of note are the howler monkeys, which make the selva resound with their morning and evening choruses. The small, agile squirrel monkey, the most ubiquitous of Amazonia's monkeys, is used in laboratories, as is the larger spider monkey. Among a host of other primate species are woolly monkeys, capuchin monkeys, titis, sakis, and marmosets. All species are used for food and frequently are seen for sale in local markets. As the human population increases and the shotgun replaces the blowgun, hunting of the wild fauna has been mounting.

Large cats, such as the jaguar and ocelot, have become rare, though pumas may be found in larger numbers in the Andean fringe of the basin. Smaller carnivores include coati, grisons, and weasels. Countless bats inhabit the Amazonian night, including the blood-eating vampire bat.

Other animals of the forest include two varieties of arboreal sloths, three types of anteaters, armadillos, and iguanas, the last especially prized for their flesh. Among snakes the nonpoisonous boa constrictor and anaconda are notable for their size, the latter reaching lengths up to 9 metres (30 feet).

The Amazon basin is exceedingly rich in birdlife. Morning and evening, the parrots and macaws fly to and from their feeding grounds, their brilliant plumage flashing in the sunlight and their raucous voices calling out their presence. Throughout the day the caciques quarrel in trees where their hanging nests swing by the dozens. Hoatzins screech in noisy flocks from streamside brush, while solitary hawks and eagles scream from tree stumps. Everywhere is heard the twittering of small birds, the sound of woodpeckers, and the guttural noises of such waterbirds as herons, cormorants, roseate spoonbills, and scarlet ibises. Parakeets, which are more common in the Amazon than sparrows are in the United States, fly around in great flocks. At dusk, toucans cry a discordant plaint from the treetops and are joined by ground-dwelling tinamous and quail. The night air is filled with the cries of various species of nightjars.

The Ituri Forest

The Ituri Forest is a dense tropical rainforest lying on the northeastern lip of the Congo River basin in the Central African nation of the Democratic Republic of the Congo. Situated between 0° and 3° N latitude and 27° and 30° E longitude, the precise geographic limits of the Ituri are poorly defined, especially along its southern and western extensions. The Ituri is bounded to the north and northeast by savanna and in the east by the fertile highlands of the Western Rift Valley, while to the south and west it is contiguous with the lowland rainforest, where its rivers drain into the Congo River. The total area of the Ituri Forest is approximately 62,900 square km (24,300 square miles). The forest, which is inhabited by both Bantu-speaking and Pygmy peoples, owes its name to the Ituri River, which flows east-west across the forest into the Aruwimi River and thence to the Congo.

Physiography

The magnificence of the tropical rainforest of the Ituri cannot help but inspire the modern-day observer with the same poetic enthusiasm displayed by the famous Welsh explorer Henry Morton Stanley when he described his crossing of the area in 1887–88. The trees of the forest range in size from small saplings just inches in diameter to gigantic hardwoods reaching to heights of 52 metres (170 feet). Like the pillars of a Gothic cathedral, these giant trees are buttressed; roots run down their sides and extend great distances across the forest floor, making the ground a labyrinth of roots that anchor the trees and grab scarce nutrients from the shallow forest soil. In places where the high canopy is nearly continuous, only small, elusive patches of sunlight reach the forest floor. The lack of light at lower levels is accentuated by the darkness of the foliage of the few shrubs and small trees that can grow under such shaded conditions. Where gaps occur in the upper canopy, herbaceous plants with long leaves resembling those of the banana plant take advantage of the available light and grow in dense stands. In many places the forest has been disturbed, either by human activity or by natural treefalls that cut large swathes through the canopy and open up the forest to the strong equatorial sun. There, the vegetation near the ground is a dense tangle of nettles, creepers, and competing species of fast-growing, short-lived trees, which make walking difficult if not

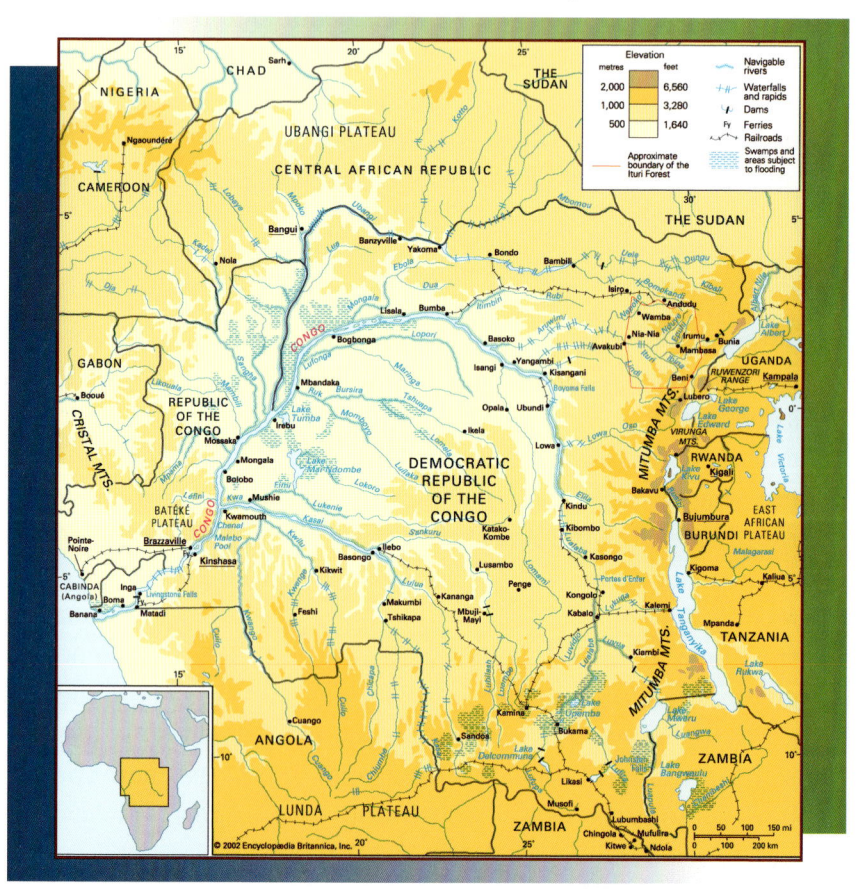

The Congo River basin and its drainage network. Encyclopædia Britannica, Inc.

impossible. Everywhere on the ground there is a profusion of fallen nuts and fruits, some as large as basketballs and many partially eaten by monkeys, antelope (duikers), and rodents. During some seasons the air is filled with the nectar of numerous species of flower, including many epiphytes, which cling to the surface of other plants and draw their sustenance from the air. Always there is the sound of myriad insects. Cicadas perch on tree trunks and emit an irritating buzz that seems designed to drive any intruder to madness. Army ants advance in columns, audibly cracking the bodies of their insect prey. Seemingly endless lines of migrating butterflies flutter through the understory and sometimes congregate in colourful displays along streambeds. The buzzing of bees, busily plying the treetops in search of sweet nourishing nectar, is ever present. While magnificent, the forest with its constant high humidity and dark interior may seem oppressive to some. Certainly Joseph Conrad thought so when he referred to the forest as the "heart of darkness." But the overwhelming impression for even the most squeamish visitor is not of darkness, not of oppressive gloom, but of life in its most vibrant and exciting form.

The Ituri Forest varies in altitude between 700 metres (2,300 feet) in its southern portions to 1,000 metres (3,300 feet) in the north. The topography is gently undulating in the south, but in the north there are frequent outcroppings of smooth granite that rise several hundred feet above the forest.

Climate and Drainage

Steeped in the tannin-rich leaves covering the forest topsoil, the water flowing in the numerous streams that drain the Ituri is the colour of strong tea. Besides the Ituri River itself, there are many broad streams that flow generally from east to west. The most notable are the Nepoko in the north, the Epulu and Nduye in the centre, and the Ibina in the south. None of these rivers is navigable, even by pirogue, for more than a few miles. The streams are fed by rains that are highly variable from month to month and from year to year. Average annual rainfall is 1,900 mm (75 inches), and there are approximately 2,000 hours of sunshine per year. Average temperature at lower elevations is 31 °C (88 °F). There is a dry season that lasts roughly from December through February, when less than 7 inches of rain normally falls. By the end of the dry season humidity within the forest is reduced, and the smaller forest streams dry up. The heaviest rains fall in October and early November; rivers overflow their banks, and large areas of the forest become flooded, making walking through the forest or driving on the few available roads extremely difficult.

Soils

The soils of the Ituri Forest developed from granites, gneisses, and metamorphosed rock of Precambrian age. In most places the soil is sandy clay or sandy clay loam, ranging in colour from reddish brown through ochre to yellowish brown and even white. The soils are acidic, and the layer of humus is thin. If exposed to the strong equatorial sun and high rainfall, as when the forest vegetation is cleared by Bantu farmers, the soil deteriorates rapidly, recovering only if it is again taken over by secondary forest. Traditionally, farmers have practiced shifting cultivation to allow the fragile soils to regenerate.

Plant and Animal Life

The climax-forest vegetation left undisturbed by human occupation is characterized by three dominant species of tall, hardwood legumes in the subfamily Caesalpinioideae. In the south and west *Gilbertiodendron deweverei* dominates and can constitute 90 percent of the standing vegetation. The regions of the forest dominated by only a few plant species have less abundant and diverse animal life than the other, more botanically mixed areas, such as in the north and east. There, *Cynometra alexandrii* and *Brachystegia laurentii*, which together comprise less than 40 percent of the canopy, are interspersed with numerous other tall species (such as *Albizia*, *Celtis*, and *Ficus*).

For many generations, people residing in the Ituri have been practicing a form of agriculture that entails clearing and burning the forest, growing their crops, and then moving after several years to allow the forest to regenerate during a long fallow period. This method of shifting slash-and-burn cultivation has created a patchwork of climax vegetation interspersed with various successional stages of secondary forest on the sites of old gardens and abandoned villages. Some areas are a tangle of lianas and shrubs beneath emerging hardwood trees, while others are in less-advanced stages of succession, with large stands of umbrella trees (*Musanga cecropoides*). These various seral patches—combined with river valleys, swampy waters, rock outcroppings, and the most recent village and garden clearings near the roads—produce a mosaic of diverse habitats that provide cover and food for the greatest abundance of mammals in forested Africa.

Situated near the forest-savanna edge, Ituri fauna include not only species typical of the African equatorial forest but also forms, such as the hyena, that are usually found on the open savanna. The most notable species is the forest giraffe, called okapi, which is endemic to the Ituri. Numerous forest antelopes include five species of duiker, the water chevrotain, and the pygmy antelope. Leopards, genets, and mongooses are the main carnivores. The elephant, buffalo, and bongo (a kind of antelope) are present in forms slightly smaller than their savanna relatives. The Ituri also supports the greatest diversity of primates of any comparable area in the world. The many monkeys include the terrestrial anubis baboon, as well as the leaf-eating imperial black and white colobus and the owl-faced monkey. The only ape is the chimpanzee. Hundreds of species of birds have been recorded; among them, the shy Congo peacock, discovered in 1936, is perhaps the most famous.

Efforts to preserve the fauna and flora are largely confined to the Maiko National Park on the southern edge of the Ituri and the Okapi Wildlife Reserve (designated a UNESCO World Heritage site in 1996) to the northeast. Both offer some protection for such animals as the forest elephant, the okapi, the Congo peacock, the aardvark, and the chimpanzee, but poaching activities and destruction of forest habitat seriously threaten these and other species both outside and inside the park. Conservation efforts have also been disrupted by civil strife that began in the late 1990s and continued into the 21st century.

OTHER FORESTS

Beyond the Amazon and Ituri forests in the tropics, a number of other forests, notable for their historical, economical, or recreational significance, occur in temperate and high-latitude regions. The forested regions listed below include the Bavarian, Black, and Belovezhskaya forests of central and eastern Europe and the Tongass and Sierra National Forests of North America.

Forests

The Bavarian Forest

The Bavarian Forest, which is known in German-speaking regions as *Bayerischer Wald*, is a mountain region in east-central Bavaria *Land* (state) in southeastern Germany. The Bavarian Forest occupies the highlands between the Danube River valley and the Bohemian Forest along Bavaria's eastern frontier with the Czech Republic. Located largely in the *Regierungsbezirk* (administrative district) of Niederbayern (Lower Bavaria), the highlands parallel the southeasterly flowing Danube for about 145 km (90 miles) from the Cham and lower Regen rivers to the Austrian border east of Passau.

The Bavarian Forest, occupying mainly granite and gneiss hills, is divided into two sections by a sharp quartz ridge known as the Pfahl. The ridge runs roughly along the Regen valley and ranges from 20 to 30 metres (65 to 100 feet) in height. The Vorderer Forest, or Danube Hills, a rolling plateau situated to the southwest between the Danube and the Pfahl, seldom rises more than 1,000 metres (3,300 feet) above sea level. Meadow, isolated farmsteads, and small hamlets dominate the landscape; only the higher and steeper slopes are still wooded. Northeast of the Pfahl is the Hinterer Forest, a higher and almost continuously forested mountain region where human settlement is confined to a few longitudinal valleys. Its highest peaks include the Grosser Arber, with an elevation of 1,456 metres (4,777 feet), and the Rachel, Lusen, Dreisesselberg, and Grosser Falkenstein.

The climate of the highlands is severe and wet, supporting only modest yields of rye, oats, and potatoes produced on small valley farms. Coniferous forest predominates, with spruce the main species at higher altitudes and a mixed woodland of spruce, silver fir, and beech found at lower levels. Lumbering, woodworking, and glass grinding are the principal industries. The tourist trade is expanding as the reputation of the Bavarian Forest as a beautiful and uncrowded holiday resort area spreads. Each year many visitors explore the Bavarian Forest National Park, where more than 98 percent of the park's 130.8-square-km (50.5-square-mile) area is tree-covered and many species of plants, birds, and small animals thrive. Principal towns of the mountain region are Regen, Zwiesel, Waldkirchen, and Grafenau.

THE BELOVEZHSKAYA FOREST

The Belovezhskaya Forest, which is also called the Belovezh Forest and Białowieża Forest, is a forested area in western Belarus and eastern Poland. It is one of the largest surviving areas of primeval mixed forest (pine, beech, oak, alder, and spruce) in Europe and occupies more than 1,200 square km (460 square miles). The Belovezhskaya Forest is located in Brest and Hrodna (Grodno) *oblasti* (provinces) of Belarus and in Podlaskie *województwo* (province) of eastern Poland near the headwaters of the Narev (Polish: Narew) and Lesnaya (Leśna) rivers, tributaries of the Bug. The forest has a wide range of flora (some conifers and hardwoods have attained ages of 350 to 600 or more years, heights in excess of 45 metres [150 feet], and diameters greater than 2 metres [6 feet]) and fauna (including elk, deer, lynx, and wild boar) from both western and eastern Europe. Hunted into extinction in the wild after World War I, the European bison, or wisent, was reintroduced to the Belovezhskaya with zoo-bred animals. The forest remains the European bison's most notable home, though the animals are now also found again in other parts of Europe, including Lithuania, Russia, and Ukraine. Once the hunting grounds of kings and tsars, the Belovezhskaya is the oldest nature preserve in Europe. Both the Polish and Belarusian portions of the forest have become national parks, and both areas were designated as World Heritage sites (the Polish portion in 1979 and the Belarusian portion in 1992).

THE BLACK FOREST

The Black Forest, which is called *Schwarzwald* in German, is a mountain region within Baden-Württemberg *Land* (state) in southwestern Germany. It is the source of the Danube and Neckar rivers. The Black Forest occupies an area of 6,009 square km (2,320 square miles) and extends toward the northeast for about 160 km (100 miles) from Säckingen on the Upper Rhine River (at the Swiss border) to Durlach (east of Karlsruhe). Its width varies from 16 to 40 km (10 to 25 miles). Structurally and topographically, it forms the counterpart of the Vosges, which lies west of the Rhine Valley. The Black Forest drops abruptly to the Rhine plain but slopes more gently toward the Neckar and Nagold valleys to the east.

It is mainly a granite highland with rounded summits, although its northern part comprises forested sandstone; and it is bordered to the south by a narrow band of lower and more fertile limestone. Divided into two parts by the deep Kinzig Valley, its highest summits—Feldberg (1,493 metres [4,897 feet]), Herzogenhorn, and Blössling—are to the south. Its northern half has an average height of 600 metres (about 2,000 feet).

The raw climate of the higher districts supports only hardy grains, but the valleys are mild with good pastureland. Oak and beech woods clothe the lower slopes, while the extensive fir forests, which gave the range its name, climb to 1,220 metres (4,000 feet). Traditional economic activities—such as lumbering, woodworking, and the manufacture of watches, clocks, and musical instruments—continue. Newer manufactures include electronic equipment and precision machinery. Tourism and winter sports are also prominent, and there are many mineral springs and spas, such as Baden-Baden and Wildbad. Principal cities are Freiburg im Breisgau, Offenburg, Rastatt, and Lahr.

Boise National Forest

This large area of evergreen coniferous forest in southwestern Idaho, U.S., is located north and east of the city of Boise. Established in 1908, it has an area of about 10,570 square km (4,080 square miles). Portions of both Frank Church–River of No Return Wilderness and Sawtooth Wilderness Area are located in the forest. Payette National Forest borders it on the north, and Challis and Sawtooth national forests adjoin it to the east.

The national forest is divided into two segments. The principal part lies east of the North Fork Payette River and is roughly teardrop-shaped, with maximum dimensions of some 175 km (110 miles) north-south and 90 km (55 miles) east-west. To the west and separated from it by the valley of the North Fork Payette River is a narrow outlying section of the forest, about 60 km (37 miles) north-south and 19 km (12 miles) east-west at its widest point; most of it, however, is much narrower. The main portion of the forest is generally steep and mountainous, the Sawtooth Range constituting much of the area; elevation decreases somewhat toward the Boise River valley in the southwest. The highest point within the forest is Big Baldy, 2,963 metres (9,722 feet) above sea level. The Middle Fork Salmon and South Fork Salmon rivers, the Middle Fork Payette River, and all three forks of the Boise River have their sources in the forest. Tree species include ponderosa and lodgepole pine, Douglas and grand fir, and Engelmann spruce. Lakes and rivers have populations of trout and salmon. Deer and elk, bobcats and lynx, wolves, bears, and otters are among the forest's mammal species.

Following the discovery of gold north of Boise in 1862, numerous mining communities sprang up in the locality. As the deposits were depleted, many of the mining sites became lumbering or livestock-grazing outposts; ponderosa pine logging and sheep ranching were important to the local economy. Some of the mining towns, however, died out and became ghost towns of interest to tourists. Logging remains an important use of the forest.

The national forest is a popular recreation area, noted for its fishing and hunting, hiking, and rafting on the Middle Fork Salmon River. The two scenic wilderness areas on the eastern side of the forest, where commercial exploitation is forbidden, are kept virtually roadless, although the Frank Church wilderness has several airstrips. Bogus Basin Ski Resort is a winter-sports centre, 26 km (16 miles) north of Boise, which is the forest headquarters.

The Franconian Forest

The Franconian Forest, which is called *Frankenwald* in German-speaking regions, is a forested highland region in extreme northeastern Bavaria *Land* (state) in east-central Germany. It forms a physical and geological link between the highlands of the Fichtel Mountains and the Thuringian Forest. About 50 km (30 miles) long, the forest descends gently north and east toward the Saale River but more precipitously west to the Bavarian Plain. Its highest point is Mount Döbra 795 metres [2,608 feet] high). Along the centre lies the watershed between the Main and the Saale basins and between the Rhine and the Elbe systems. The principal tributaries of the Main rising in the forest are the Rodach and the Hasslach; of the Saale, the Selbitz.

Small hamlets lie in clearings in the heath, bog, and woods of the Franconian Forest. The chief city is Hof, to the east. Kulmbach, Kronach, and Bayreuth lie to the west.

Mount Hood National Forest

The Mount Hood National Forest is a mountainous, heavily forested region in northwestern Oregon, U.S. The forest starts about 32 km (20 miles) east of Portland and extends southward along the Cascade Range from the Columbia River for more than 100 km (60 miles). Established in 1908 as the Oregon National Forest, it covers some 4,850 square km (1,875 square miles) of scenic mountains and streams. The forest provides timber, water, forage, wildlife habitats, and recreation; it is drained by the Columbia, Sandy, Clackamas, Hood, and White rivers and their tributaries. Douglas fir is the dominant tree

Punch Bowl Falls, Mount Hood National Forest, Oregon. B. Nelson—Shostal

species. Mount Hood (3,424 metres [11,235 feet]), near the centre of the forest, is Oregon's highest point.

Features of the national forest include Mount Hood Wilderness Area and four smaller wilderness areas, Timberline Lodge (built in 1937 on Mount Hood), Multnomah Falls (190 metres [620 feet]), Austin and Bagby hot springs, Timothy Lake, portions of the Oregon Trail, and Eagle Creek Trail, leading through a region of waterfalls. Hiking and skiing are two of the many activities available to tourists. Pacific Crest National Scenic Trail traverses the forest from north to south. The Warm Springs reservation of the Paiute, Wasco, and Warm Springs Indians adjoins the forest on the southeast, and the Columbia River Gorge National Scenic Area runs along its northern border. Headquarters are at Sandy.

Ozark–Saint Francis National Forest

The Ozark–Saint Francis National Forest is made up of forested areas mainly located in central and northwestern Arkansas, U.S., but it also includes a segment along the Mississippi River in the eastern part of the state. The forest consists of several separate units embracing parts of the Ouachita and Boston mountains and the southern end of the Springfield Plateau. The westernmost part of Ozark National Forest reaches the Oklahoma border, while St. Francis National Forest constitutes the eastern segment.

Established in 1908, Ozark National Forest covers approximately 4,048 square km (1,563 square miles). It is drained by tributaries of the Arkansas River. Hardwood trees, mainly oak and hickory, form the primary vegetation, and the undergrowth includes dogwood, maple, and redbud. Animal life is plentiful and includes white-tailed deer, black bear, rabbits, bobwhite quail, and wild turkey. The main unit constitutes a roughly rectangular area extending west-east through the southern Boston Mountains and surrounded by smaller units; the park Buffalo National River lies north of its eastern half. There are five designated wilderness areas, four in the main unit (Upper Buffalo, Hurricane Creek, Richland Creek, and East Fork) and one (Leatherwood) in the smaller unit northeast of the main unit. In the northernmost section of the forest, 23 km (14 miles) northwest of the city of Mountain View, is Blanchard Springs Caverns, a three-level cave system with two levels accessible to visitors. Magazine Mountain, at an elevation of 839 metres (2,753 feet), the highest point in Arkansas, is located in the smaller unit south of the main part.

Saint Francis National Forest, established in 1960 and administered jointly with Ozark National Forest, consists of 85 square km (33 square miles) of bottomland hardwood trees. It is named for the St. Francis River, which, along with the Mississippi River, forms the forest's eastern boundary. The northwestern portion of the forest is located on hilly Crowley's Ridge. Popular fishing areas and hiking trails are found in and around Storm Creek Lake and Bear Creek Lake.

Six national scenic byways cross the forests, including Arkansas Highway 7, often considered the most beautiful drive in the state. The 370 km (230 miles) of hiking trails include the 266-km (165-mile) Ozark Highlands National Recreational Trail, which begins in the southwest corner of the main unit at Lake Fort Smith, winds eastward, and terminates in the northeastern corner at Buffalo River Trail in Buffalo National River. Six waterways within the forests are designated as national wild and scenic rivers.

Sequoia National Forest and Giant Sequoia National Monument

The Sequoia National Forest and Giant Sequoia National Monument is a large natural region of mountains and forestland in east-central California, U.S. The area is noted for its more than three dozen groves of big trees, or giant sequoias (*Sequoiadendron giganteum*), for which the national forest and the national monument are named.

The region lies at the southern end of the Sierra Nevada and is divided into two main sections of unequal size. The smaller, northern portion is separated from Sierra National Forest (north) by the Kings River and adjoins Kings Canyon (east) and Sequoia (south) national parks. The larger, southern section borders Sequoia National Park (north) and Inyo National Forest (northeast and east) and stretches south nearly to Bakersfield and the Mojave Desert (south). Two small, separate units of Sequoia National Forest lie immediately south and southeast of this section, and the Tule Indian Reservation adjoins its western border. The national forest, which was established in 1908, has a total area of 4,610 square km (1,780 square miles), with elevations ranging from 300 to above 3,700 metres (1,000 to above 12,000 feet). Giant Sequoia National Monument was established within Sequoia National Forest in 2000 to protect the groves of giant sequoias dotted throughout both sections of the forest. The national monument occupies a total of some 1,329 square km (513 square miles) of land and includes campgrounds and an interpretive trail.

The forest, which includes conifers, hardwoods, and chaparral as well as big trees, provides timber, water, forage, wildlife habitat, and recreation. In addition to the groves of giant sequoias, notable features include Kern River and Kings River canyons, Boyden Cavern, Balch Park (which has a notable grove of redwoods), many mountain lakes and well-stocked trout streams, and the Boole Tree, with a height of 82 metres (269 feet) and a circumference of 11 metres (35 feet), the largest known tree in any U.S. national forest. Dome Land Wilderness, one of five wilderness areas within the national forest, is a lofty region northeast of Bakersfield containing numerous rock outcroppings.

Sequoia National Forest is a popular hunting area, particularly for such large game as bears and mule deer. Trout fishing, camping, skiing, snowmobiling, and kayaking are also widely enjoyed, and the Kings and North Fork Kern rivers are major destinations for white-water rafting trips. The 1,450 km (900 miles) of trails are used by hikers, horseback riders, and off-road vehicle operators. The Pacific Crest National Scenic Trail runs through parts of the forest. Headquarters are at Porterville.

Sherwood Forest

Sherwood Forest is a woodland and former royal hunting ground, county of Nottinghamshire, England, that is well known for its association with Robin Hood, the outlaw hero of medieval legend. Sherwood Forest formerly occupied almost all of western Nottinghamshire and extended into Derbyshire. Today a reduced area of woodland, mostly pine plantations, remains between Nottingham and Worksop. In the north the great ducal estates, or "dukeries," of Welbeck, Clumber, and Thoresby have preserved parts of the forest. Many veteran oaks remain, and there is much heath. Agricultural encroachment has been limited by the poor, sandy soil. An underlying coalfield has been extensively developed since the mid-19th century.

SIERRA NATIONAL FOREST

The Sierra National Forest encompasses a region of forests and streams in central California, U.S., extending along the Sierra Nevada between Yosemite and Kings Canyon national parks (north and southeast, respectively) and bordered by Inyo (northeast), Sequoia (south), and Stanislaus (northwest) national forests. It was established in 1905 from an earlier (1893) forest reserve. It has an area of about 5,260 square km (2,030 square miles) and elevations ranging from 275 to 4,000 metres (900 to 13,000 feet). Notable features include the big trees (giant sequoias) of the Nelder and McKinley groves and the Kaiser, Dinkey Lakes, Monarch, Ansel Adams, and John Muir wilderness areas (the last two shared with Inyo National Forest). Headquarters are at Clovis.

Vegetation ranges from alpine meadows to stands of lodgepole pines, red and white firs, cedars, mountain hemlocks, and aspen. The forest and its streams provide timber, grazing land, water, and hydroelectric power, and there is some gold mining. Wildlife includes mule deer, black bears, coyotes, bobcats, foxes, marmots, porcupines, and quail. Trout, bass, and bluegills are favourite sportfishing species. The Mount Dana–Minarets Escarpment is one of the forest's scenic highlights, many of its peaks exceeding 3,700 metres (12,000 feet). Hundreds of miles of trails are maintained in the Mammoth–High Sierra area. The highest peak, Mount Humphreys (4,263 metres [13,986 feet]), is on the Sierra Nevada crest. The Pacific Crest National Scenic Trail crosses the forest, and Devils Postpile National Monument is near its eastern edge. The Sierra Vista Scenic Byway runs for 160 km (100 miles) past some of the area's natural highlights. The region offers fishing in numerous cold, clear streams and lakes, as well as bear, deer, and quail hunting. Winter sports such as snowmobiling and skiing are popular, as are aquatic sports such as kayaking and white-water rafting in warm weather.

The Teutoburg Forest

The Teutoburg Forest, or *Teutoburger Wald* in German, is the westernmost escarpment of the Weser Hills (Weserbergland) in northeastern North Rhine-Westphalia *Land* (state) in northern Germany. Its wooded limestone and sandstone ridges curve from the Ems River valley southeastward in an arc approximately 100 km (60 miles) long and 6.5 to 9.5 km (4 to 6 miles) wide around the north and northeast sides of the Münsterland basin. The highest point in the Teutoburg Forest, the Velmerstot, rises to an elevation of 468 metres (1,535 feet) at the southeastern end where the range meets the Egge Mountains. The city of Bielefeld, a diversified industrial centre most famous for its linen textiles, is situated at an important pass through the hills. The Hermannsdenkmal, a colossal metal statue built in the 19th century to commemorate the Battle of the Teutoburg Forest (fought 9 CE), in which Germanic tribes led by Arminius (German: Hermann) annihilated three Roman legions, stands outside Detmold on the northeastern slope. Numerous health and holiday resorts are established in the small hill towns situated among beech and spruce forests.

The Thuringian Forest

The Thuringian Forest, known as *Thüringerwald* in German-speaking regions, is a range of forested hills and mountains in Germany, extending in an irregular line from the neighbourhood of Eisenach in west-central Thuringia southeastward to the Bavarian frontier, where it merges with the Franconian Forest. Its breadth varies from 10 to 35 km (6 to 22 miles). It nowhere rises into peaks, and, of its rounded summits, the highest, Beerberg, rises only 982 metres (3,222 feet). This range encloses many charming valleys and glens; the most prominent feature of its scenery is formed by the forests, chiefly of pines and firs. The northwest part of the system is the loftier and the more densely wooded as well as the more beautiful. The southeast part is the more populous and industrial.

The crest of the Thuringian Forest, from the Werra to the Saale rivers, is traversed by the Rennsteig, a broad path of unknown antiquity. The name probably means "frontier path," and the path marked in fact the historical boundary between Thuringia and Franconia. It may be also regarded as part of the boundary line between north and south Germany, for dialect, customs, local names, and native costume have traditionally been different on the two sides. The area was once an iron-mining centre (until the 16th century) but is now largely given over to small industries (toy making, wood carving, and glass and china manufacturing) and to tourism in such resorts as Eisenach, Friedrichroda, Giessübel, and Oberhof.

Tongass National Forest

The Tongass National Forest is a forest region and wilderness area in southeastern Alaska, U.S. It was established in 1907 by an executive order issued by President Theodore Roosevelt (formal legislation declaring it a national forest was signed into law in 1909). Tongass National Forest covers most of the Alaska panhandle and is the largest publicly owned forest in the United States. The forest was named for a Tlingit Indian group. Its approximately 68,790 square km (26,560 square miles), composed of (volcanic) mountainous offshore archipelagos and rugged fjord-indented coastline, include some of the most extensive intact remnants of virgin temperate rainforest in North America. Misty Fjords (or

Fiords) and Admiralty Island national monuments are located within the forest.

Found in association with barren alpine tundra (above the tree line), lowland muskeg, and scores of tidewater glaciers that descend to the coastline, the forest is dominated by towering species of western hemlock and Sitka spruce. Its rich understory includes blueberries, skunk cabbages, and a profusion of ferns and mosses. Brown and black bears, Sitka black-tailed deer, wolves, mountain goats, river otters, mink, northern flying squirrels, seals, and numerous species of birds—including bald eagles, northern goshawks, and the elusive marbled murrelet—are part of the unusual variety of wildlife. Approximately one-third of Tongass has been protected as a national wilderness area, and about one-fifth has been designated for commercial development. The forest has long been the centre of intense conflict between conservationists and logging companies. Fishing and tourism are the forest's most significant economic activities.

CHAPTER 2

Forestry

Forestry is defined as the management of forested land, together with associated waters and wasteland, primarily for harvesting timber. To a large degree, modern forestry has evolved in parallel with the movement to conserve natural resources. As a consequence, professional foresters have increasingly become involved in activities related to the conservation of soil, water, and wildlife resources and to recreation.

This chapter traces the history of forestry from its origin in ancient practices to its development as a scientific profession in the modern world, and it discusses the kinds and distribution of forests as well as the principal techniques and methods of modern forest management in detail.

THE HISTORY OF FORESTRY

Because of their ecological and economic value, humans have developed a special kinship with forests. Forests have acted as civilization's raw materials storehouses since even before the dawn of modern humans. Forests are home to many game animals, medicinal and agricultural plants, and woods useful for heating, cooking, and construction. Since these biological systems are so valuable, many cultures and civilizations have developed plans and rules to ensure their conservation, especially as human populations increased.

THE ANCIENT WORLD

It is believed that *Homo erectus* used wood for fire at least 750,000 years ago. The oldest evidence of the use of wood for construction, found at the Kalambo Falls site in Tanzania, dates from some 60,000 years ago. Early organized communities were located along waterways that flowed through the arid regions of India, Pakistan, Egypt, and Mesopotamia, where scattered trees along riverbanks were used much as they are today—for fuel, construction, and handles for tools. Writers of the Old Testament make frequent reference to the use of wood. Pictures in Egyptian tombs show the use of the wooden plow and other wooden tools to prepare the land for sowing. Carpenters and shipwrights fabricated wooden boats as early as 2700 BCE. Theophrastus, Varro, Pliny, Cato, and Virgil wrote extensively on the subject of trees, their classification, manner of growth, and the environmental characteristics that affect them.

The Romans took a keen interest in trees and imported tree seedlings throughout the Mediterranean region and Germany, establishing groves comparable to those in Carthage, Lebanon, and elsewhere. The fall of the Roman Empire signaled an end to conservation works throughout the Mediterranean and a renewal of unregulated cutting, fire, and grazing of sheep and goats, which resulted in the destruction of the forests. This, in turn, caused serious soil loss, silting of streams and harbours, and the conversion of forest to a scrubby brush cover known as maquis.

Medieval Europe

In medieval Europe, forest laws were aimed initially at protecting game and defining rights and responsibilities. Hunting rights were vested in the feudal lord who owned the property and who had the sole right to cut trees and export timber. Peasants were permitted to gather fuel, timber, and litter for use on their own properties and to pasture defined numbers of animals. By 1165, however, land clearing for agriculture had gone so far that Germany forbade further forest removal. The systematic management of forests had its true beginnings, however, in the German states during the 16th century. Each forest property was divided into sections for timber harvesting and regeneration to ensure a sustainable yield of timber for the entire property. This working plan called for accurate maps and assessments of timber volume and expected growth rates.

Trees have been raised from seed or cuttings since biblical times, but the earliest record of a planned forest nursery is that of William Blair, cellarer to the Abbey of Coupar Angus in Scotland, who raised trees to grow in the Highland Forest of Ferter as early as 1460. After the dissolution of the monasteries, many newly rich landowners in Scotland and England found a profitable long-term investment in artificial plantations established on poor land. John Evelyn, a courtier in the reign of Charles II, published his classic textbook *Sylva* in 1664, exhorting them to do so, and today virtually all of Britain's 2.1 million hectares (5.2 million acres) of woodland consists of artificial plantations. Other countries had managed their natural forests better and had little need, until recently, to afforest (or plant trees to create forests) on bare land. The 20th century, however, has seen a tremendous expansion of artificial plantations in all the continents, planned to meet the ever-growing needs for wood and paper as essential materials in modern civilization.

Modern Developments

Formal education in forestry began about 1825 when private forestry schools were established. These were the outgrowth of the old master schools such as Cotta Master

School, which developed into the forestry college at Tharandt—one of the leading forestry schools in Germany. The National School of Forestry was established in Nancy, France, in 1825.

During the 19th century the reputation of German foresters stood so high that they were employed in most continental European countries. Early American foresters, including the great conservation pioneer Gifford Pinchot, gained their training at European centres. But the doctrine of responsible control had to fight a hard battle against timber merchants who sought quick profits.

The 20th century has seen the steady growth of national forest laws and policies designed to protect woodlands as enduring assets. Beginning in the 1940s vast land reclamation was undertaken by Greece, Israel, Italy, Spain, and the Maghrib countries of North Africa to restore forests to the slopes laid bare by past abuse. The main objective of the tree planting is to save what remains of the soil and to protect the watersheds. In China, where forests once extended over 30 percent of the land, centuries of overcutting, overgrazing, and fires reduced this proportion to approximately 7 percent. China has taken major steps to improve land use, including construction of reservoirs and a huge forest planting program, which reported the planting of 15.75 million hectares (38.92 million acres) between 1950 and 1957 alone.

The character of forest policies around the world reflects national political philosophies. In Communist countries all forests are owned by the state. In the United States both the federal and the state governments have deemed it prudent to hold substantial areas of natural forest, while allowing commercial companies and private individuals to own other areas outright. Similar patterns of ownership are found throughout most of Asia, western Europe, and the Commonwealth countries. In Japan the

A child earths up a tree with his parents during National Tree Planting Day in Jiangsu Province, China. China Photos/Getty Images

extensive forests are largely state owned. Tribal ownership is found in many African countries and proves a serious obstacle to effective modern management. International cooperation is effected by the Forestry Department of the United Nations' Food and Agriculture Organization, with headquarters in Rome.

THE DEVELOPMENT OF U.S. POLICIES

The history of forestry in the United States followed the same path as forestry in Europe—land clearing, repeated burning, overcutting, and overgrazing—until a bill was passed by Congress in 1891 authorizing the president to set apart from the public domain reserves of forested land. In 1905 an act of Congress, with strong encouragement from President Theodore Roosevelt, transferred the

Bureau of Forestry from the Department of the Interior to the Department of Agriculture. Gifford Pinchot, who had been chief of the bureau, was made chief of the newly named Forest Service. Pinchot developed the U.S. Forest Service into a federal agency that today is recognized worldwide for its research, education, and land and forest management expertise. On the state level the Morrill Act of 1862 provided for federal–state cooperative programs in which the federal government granted first land, then money, to the states for the establishment of technical agricultural colleges. The Weeks Act of 1911 authorized the federal government to assist the states in protecting forests from fire, and the Clark–McNary Act of 1924 extended the provisions of the Weeks Act to include cooperation in forest extension, planting, and assistance to forest owners. During the Great Depression of the 1930s the interests of forestry were served most imaginatively and thoroughly by the Civilian Conservation Corps (CCC), which planted trees, fought forest fires, and improved access to woodlands across the United States. The CCC, rooted in the system of public works initiated by President Franklin D. Roosevelt, continued until 1942, acquainting many people with forestry as a major government activity.

The complete mobilization of resources for the U.S. involvement in World Wars I and II and the pent-up demand for consumer goods made heavy demands on forest resources and industries. As a result, forestry on a national basis entered a period of the most rapid advance since the turn of the century. This time the advance was stimulated by the need for forest products and by the conviction on the part of the major timber companies that they must protect their raw material supply. To protect the forests from growing pressure from single-interest groups, Congress passed the Multiple Use–Sustained Yield Act of

Men from the Reforestation Army, part of the Civilian Conservation Corps, clear brush from and plant seedlings on a hillside in the St. Joe National Forest of Idaho in the 1930s. FPG/Archive Photos/Getty Images

1960. This act directed that the national forests be managed under principles of multiple use so as to produce a sustained yield of products and services. The Bureau of Outdoor Recreation was established shortly thereafter in the Department of the Interior. The Land and Water Conservation Fund, established in 1964, launched a comprehensive program for planning and developing outdoor recreation facilities. State forestry programs had their beginnings in the United States during colonial times, but it was the Weeks and Clark–McNary laws that provided the impetus to develop recognized state forestry departments. The Smith–Lever Act of 1914 allotted funds through the state agricultural colleges for extension work in forestry. Initial programs emphasized tree planting and demonstrations, but today all aspects of forestry and natural and related resources are included.

Industrial forestry began around 1912 when Finch, Pruyn, and Company started a forestry program on its Adirondack holding in New York. Trees to be cut were marked by foresters, and the cutting budget was projected on a sustained-yield basis. A rapid expansion of company forestry programs in the northeastern United States began in the late 1920s and early 1930s. Following World War II, paper companies expanded rapidly throughout the South and West and to a lesser extent in the Northeast. Pulp and paper companies were quick to recognize the benefits to be realized from research financed by the Forest Service and by universities in such fields as tree physiology, entomology, genetics, and tree improvement. A few companies established their own experimental forests and research teams.

The cause of forestry in the United States also has been advanced by citizens' organizations. These vary from lay and youth organizations, such as the Boy Scouts and garden clubs, to the nation's most prestigious scientific societies. The American Association for the Advancement of Science stimulated Congress in 1876 to embark on a sustained federal forestry program. The National Academy of Sciences 1896 report on forest reserves began its long involvement in forest conservation. The Society of American Foresters, founded in 1900, together with its sister societies in Canada and Mexico, represents the profession of forestry in North America.

THE CLASSIFICATION AND DISTRIBUTION OF FORESTS

Botanical classification places forest trees into two main groups: Gymnospermae and Angiospermae. The gymnosperms consist exclusively of trees and woody shrubs, whereas the angiosperms are a diverse group of plants that include trees and shrubs as well as grasses and herbaceous plants. The gymnosperms probably gave rise to the angiosperms, although the manner in which this took place is disputed.

Gymnosperms

The gymnosperms are of very ancient lineage and include the earliest trees on the evolutionary scale. With certain exceptions, the seeds of gymnosperms are borne in cones, where they develop naked or exposed on the upper surface of the cone scales. The wood of these trees has a simple structure. Many species are extinct, such as the tree ferns of the Carboniferous Period (345 to 280 million years ago), and are known only as fossils. The ginkgo, or maidenhair tree, is the sole survivor of an entire order of gymnosperms, the Ginkgoales. Among the gymnosperms, the most important and numerous forest trees are the conifers, also known as softwoods. This group includes the well-known pines, spruces, firs, cedars, junipers, hemlocks, and sequoias. These species are so dominant in the gymnosperm class that forests of gymnosperm trees are typically called coniferous forests. Except for the ginkgo, larches, and bald cypress, all gymnosperms are evergreen.

Angiosperms

The angiosperms constitute the dominant plant life of the present geologic era. They are the products of a long line of evolutionary development that has culminated in the highly specialized organ of reproduction known as the flower, in which seed development occurs within an ovary. This group includes a large variety of broad-leaved trees, most with a deciduous leaf habit but some that are evergreen. The angiosperms are further divided into monocots and dicots. Trees are represented in both groups.

Monocots

The monocots include principally the palms and bamboos. Palm trees form extensive savannas in certain tropical and subtropical zones but are more usually seen along watersides or in plantations.

Palm trees have no growth rings, being made up of spirally arranged bundles of fibres, giving a light, spongy wood. Palms are valuable, however, for their various fruits (coconuts, dates, and palm kernels) and leaf products (carnauba wax, raffia, and thatching and walling materials for houses in the tropics).

Another form of tropical monocotyledonous forest is the bamboo thicket, common in Asia, composed of giant woody grasses. One of the most versatile plants in the world, bamboo is valuable as a construction material, as well as for hundreds of other applications. Its young shoots are eaten as vegetables and are a valuable source of certain enzymes.

Dicots

Finally, a more highly evolved group of forest trees is the dicots, or broad-leaved trees, also called hardwoods. Their wood structure is complex, and each sort of broad-leaved lumber has characteristic properties that fit it for particular uses.

The Occurrence and Distribution of Forests

Approximately 4 billion hectares (9.9 billion acres), or about one-third of the total land area in the world, is covered with closed forests of broad-leaved and coniferous species and open forests or savannas. Because of the varying characteristics of individual tree species, the kind and distribution of

DISTRIBUTION OF THE WORLD'S FOREST LAND*

REGION	TOTAL LAND AREA	CLOSED FOREST BROAD-LEAVED	CLOSED FOREST CONIFEROUS	OPEN FOREST	TOTAL FOREST AREA	PERCENT OF TOTAL LAND AREA FORESTED
North America	1,835	168	301	215	684	37
Europe	472	65	88	21	174	37
Former Soviet Union	2,227	147	645	128	920	41
Africa	2,966	216	2	500	718	24
Latin America	2,054	666	26	250	942	46
Asia	2,573	414	55	98	567	22
Pacific Area	950	50	22	70	142	15
World Totals	13,077	1,726	1,139	1,282	4,147	32

*In millions of hectares.

the world's forests are largely determined by local conditions. Each combination of temperature, rainfall, and soil has a peculiar association of trees and other vegetation that are best equipped to compete with other plants for that site. The open forest is characteristically a tropical grassland, often disturbed by fire, with forest along streams and scattered individual trees or small groves. Closed thorn forests usually appear adjacent to the savannas. In general, coniferous forests are found in the cooler, drier areas, and the broad-leaved species are predominant in the warmer, usually moister parts of the world. Tropical forests consist almost exclusively of broad-leaved species. Mixed broad-leaved and coniferous forests are found near the boundaries between these two climatic zones.

Coniferous forests are largely found in the temperate climate of the Northern Hemisphere, where they cover approximately 1.1 billion hectares (2.7 billion acres); some 85 percent of them are in North America and the erstwhile Soviet Union. The northern coniferous forest, or taiga, extends across North America from the Pacific to the Atlantic, across northern Europe through Scandinavia and Russia, and across Asia through Siberia to Mongolia, northern China, and northern Japan. It has outliers along all the temperate mountain ranges, including the Rockies, the Appalachians, the Alps, the Urals, and the Himalayas. Its principal trees are spruces (of the genus *Picea*), northern pines (*Pinus*), silver firs (*Abies*), Douglas firs (*Pseudo tsuga*), hemlocks (*Tsuga*), and larches (*Larix*). Together these northern softwood forests form a world resource of tremendous importance, yielding the bulk of the lumber and pulpwood handled commercially. Northern conifers from many lands are extensively planted in Europe, including the British Isles.

The southern coniferous forest has a discontinuous spread through the southern part of the Northern Hemisphere, including California, the southeastern states of the United States, the Mediterranean lands of southern Europe, North Africa, Asia Minor, parts of the Asian mainland, and southern Japan. Pines are the principal trees, along with cypresses (*Cupressus* and *Chamaecyparis*), cedars (*Cedrus*), and redwoods and mammoth trees (*Sequoia* and *Sequoiadendron*). Certain southern pines such as the California Monterey pine (*Pinus radiata*) grow poorly in their native habitat but exceptionally fast when planted in subtropical Europe, Africa, New Zealand, and Australia.

In addition to the plantations of introduced pines, small areas of coniferous forest are found in the Southern Hemisphere, notably the Chile pine, *Araucaria araucana*, in the Andes; hoop pine, or bunyabunya, *Araucaria bidwillii*, in Australia; and kauri pine, *Agathis australis*, in New Zealand.

The dicotyledonous broad-leaved species form three characteristic types of forests: temperate deciduous, subtropical evergreen, and tropical evergreen.

Temperate deciduous broad-leaved forests are made up of the summer-green trees of North America, northern Europe, and the temperate regions of Asia and South America. Characteristic trees are oaks (*Quercus* species), beeches (*Fagus* and *Nothofagus*), ash trees (*Fraxinus*), birches (*Betula*), elms (*Ulmus*), alders (*Alnus*), and sweet chestnuts (*Castanea*). Temperate broad-leaved trees expand their foliage in spring, grow rapidly in summer, and shed all their leaves each fall.

Subtropical evergreen broad-leaved forests grow largely in countries with a Mediterranean type of climate—that is, hot, dry summers and cool, moist winters. Their trees have characteristic thick, hard-surfaced, leathery-textured leaves with waxy coatings that enable them to resist water loss during summer droughts. Their evergreen habit enables them to make use of moist winters. Typical trees are the evergreen oaks, species of *Quercus*, and the madrone, or *Arbutus*, while in Australia most evergreen broadleaf trees are species of *Eucalyptus*. Few evergreen broadleaf trees have high timber value, and many are little more than scrub, highly inflammable during hot, dry summers. Their world distribution embraces California; the southeastern states of the United States; Mexico; parts of Chile and Argentina; the Mediterranean shores of Europe, Asia, and North Africa; South Africa; and most of Australia.

Tropical evergreen broad-leaved forests, or tropical rainforests, grow in the hot, humid belt of high rainfall that follows the Equator around the globe. They occur in West and Central Africa, South Asia, the northern zone of Australia, and in Central and South America. Where they extend into regions of seasonal rainfall, such as monsoon zones, they become less truly evergreen, holding many trees that stand leafless during the short dry seasons. Tropical rainforests hold a great variety of tree species. A few of the timbers, such as teak (*Tectona grandis*), in India, and mahogany (*Swietenia macrophylla*), in Central America, have uniquely useful properties or ornamental appearance and hence a high commercial value. Balsa, *Ochroma pyramidale*, from Central America, is the lightest timber known; it is used for rafts, aircraft construction, and insulation against noise, heat, and cold.

Trees outside areas classified as forestland, such as those in windbreaks, along rights-of-way, or around farm fields, are also important resources, especially in densely populated areas. For example, some 20 percent of Rwanda's farmland is maintained by farmers as woodlots and wooded pastures. These roughly 200,000 hectares (494,210 acres) of dispersed trees exceed the combined area of the country's natural forests and state and communal plantations. In the Kakamega District of Kenya more than 90 percent of the farms have scattered trees maintained for animal fodder and fuelwood. Of the 7.2 billion trees planted in the densely settled plains region of China, 5.8 billion have been planted around homes and in villages, with each household tending an average of 74 trees. Even in France, where trees are not used much for fuelwood, trees outside the forests occupy 883,000 hectares (2.2 million acres). There are no good estimates of the worldwide totals of such scattered trees, but their existence provides many locally useful products and extends the resources in the forested areas.

THE PURPOSES AND TECHNIQUES OF FOREST MANAGEMENT

Forest management is often complex and challenging. Forest managers must balance the desires of private and public interests, ensuring that harvesting does not threaten the sustainable yield of wood and other products. Forest management includes the maintenance of the health of the forest as a whole through disease and pest monitoring, the regulation of hunting and other recreational activities, and the replanting of clear-cut or thinned areas. Other aspects of forest management include forest fire risk reduction, erosion control, and watershed management. In some parts of the world, agroforestry and urban forestry have grown in importance as well.

Multiple-Use Concept

The forests of the world provide numerous amenities in addition to being a source of wood products. The various public, industrial, and private owners of forestland may have quite different objectives for the forest resources they control. Industrial and private owners may be most interested in producing a harvestable product for a processing mill. However, they also may want other benefits, such as forage for grazing animals, watershed protection, recreational use, and wildlife habitat. On public lands the multiple-use land management concept has become the guiding principle for enlightened foresters. This is a complex ecological and sociological concept in contrast to the single-use principle of the past. The challenge, in the words of Gifford Pinchot, is to "ensure the greatest good for the most people over the long run." Thus timber production may have top priority in some areas, but in others, such as those near large population centres, recreational values may have high priority. Multiple use calls for exceptional skill on the part of forest managers.

Sustained Yield

Forest management originated in the desire of the large central European landowners to secure dependable income to maintain their castles and retinues of servants. Today forest management is still primarily economic in essence, because modern forest industries, mainly sawmilling and paper manufacture, can be efficient only on a continuous-operation basis.

Foresters think in long time scales, in line with the long life of their renewable crop. However, it is possible that a forest can be managed in such a way that a modest timber crop may be harvested indefinitely year after year if annual harvest and the losses due to fire, insects, diseases, and other

Deciduous forest of beech in autumn, New Forest, southern England, U.K. Heather Angel/Biofotos

destructive agents are counterbalanced by annual growth. This is the sustained-yield concept. An important element is the rotation, or age to which each crop can be grown before it is succeeded by the next one. Examples of short rotation periods in the subtropics are seven years for leucaena for fuelwood, 10 years for eucalyptus, and 20 years for pine for pulpwood. Here a sustained yield could in theory be obtained simply by felling one-tenth of the eucalyptus forest each year and replanting it. Rotation periods for pulpwood in northern Europe and North America extend to 50 years. Softwood sawlogs often need 100 years to reach an economic size, while rotation periods for broad-leaved trees, such as oak and beech, in central Europe, may extend to two centuries. Over so long a growing spell only part of the lumber yield is obtained by the clear-cutting of a small fraction of the forest each year. The rest is secured by systematically thinning out the whole forest periodically.

Sustained-yield principles are likewise applied to minor forest produce. Turpentine and pitch, also known as naval stores because these resinous items were used to caulk wooden ships and thus keep them from leaking, are obtained by the systematic tapping of the lower trunk of certain subtropical pines. Successive cuts with a chisel-like tool every few days during a succession of summers eventually kill the trees. To ensure continued yields, crops of young pines are raised rotationally to replace those felled. A similar system is followed for Para rubber, *Hevea brasiliensis*, grown in plantations.

Forest Products

The culture of trees in natural forests and plantations for the yield of lumber, pulp, chips, and specialty products is a principal management objective. In many parts of the world the harvest of wood for firewood and charcoal is the

A hauling trailer pulls up to be loaded with freshly cut logs in Alabama. Washington Post/Getty Images

dominant use, and these products are often in short supply. Timber stands must be felled and regenerated in an orderly sequence to meet continuing industrial demands.

Silviculture

Silviculture is the branch of forestry concerned with the theory and practice of controlling forest establishment, composition, and growth. Like forestry itself, silviculture is an applied science that rests ultimately upon the more fundamental natural and social sciences. The immediate foundation of silviculture in the natural sciences is the field of silvics, which deals with the laws underlying the growth and development of single trees and of the forest as a biologic unit. Growth, in turn, depends on local soils and climate, competition from other vegetation, and interrelations with animals, insects, and other organisms, both beneficial and destructive. The efficient practice of silviculture demands knowledge of such fields as ecology, plant physiology, entomology, and soil science and is concerned with the economic as well as the biologic aspects of forestry. The implicit objective of forestry is to make the forest useful to man.

The practice of silviculture is divided into three areas: methods of reproduction, intermediate cuttings, and protection. In every forest the time comes when it is desirable to harvest a portion of the timber and to replace the trees removed with others of a new generation. The act of replacing old trees, either naturally or artificially, is called regeneration or reproduction, and these two terms also refer to the new growth that develops. The period of regeneration begins when preparatory measures are initiated and does not end until young trees have become established in acceptable numbers and are fully adjusted to the new environment. The rotation is the period during which a single crop or generation is allowed to grow.

Intermediate cuttings are various types of cuttings made during the development of the forest—that is, from the reproduction stage to maturity. These cuttings or thinnings are made to improve the existing stand of trees, to regulate growth, and to provide early financial returns, without any effort directed at regeneration. Intermediate cuttings are aimed primarily at controlling growth through adjustments in stand density, the regulation of species composition, and selection of individuals that will make up the harvest trees. Protection of the stand against fire, insects, fungi, animals, and atmospheric disturbances is as much a part of silviculture as is harvesting, regenerating, and tending the forest crop.

Silvicultural systems are divided into those employing natural regeneration, whereby tree crops are renewed by natural seeding or occasionally sprout regrowth, and those involving artificial regeneration, whereby trees are raised from seed or cuttings. Natural regeneration is easier but may be slow and irregular; it can only renew existing forests with the same sorts of trees that grew before. Artificial regeneration needs more effort, yet can prove quicker, more even, and in the long run more economical. It permits the introduction of new sorts of trees or better strains of the preexisting ones.

Natural Regeneration

In established forests the selective cutting of marketable timber, taking either one tree at a time (single-tree selection) or a number of trees in a cluster (group selection) and leaving gaps in which replacements can grow up from natural seedlings, can prove economical and also ensure the best possible use of available soil, light, and growing space. The best examples of single-tree-selection forests are found in Switzerland, on slopes where any clear felling could lead quickly to soil erosion and avalanches.

Alternative methods of natural regeneration deal with areas of land as units, rather than with single trees. One highly effective example is employed in the Douglas fir forests along the Pacific slope of Canada and the western United States. Logging by powerful yarding machines, using overhead cables, creates wedge-shaped gaps of cleared ground. The surrounding forest is left standing for many years in order to provide shelter and seed. Abundant seed is carried by wind on to the cleared land and gives rise, in a few years, to a full crop of seedling firs. After these have reached seed-bearing age, the areas previously left standing may be removed in their turn. Similar systems using a pattern of strips cut across the forest, or circular plots gradually extended until they meet and coalesce, are employed in France and Germany.

A silvicultural system employing practices of short rotation (five to 10 years) and intensive culture (fertilization, weed, and insect and disease control) with superior genotypes relies on coppice, or regeneration from sprouts arising from stumps of felled trees, as the method of regeneration of the new crop and is characterized by high productivity.

Artificial Regeneration

Artificial regeneration is accomplished by the planting of seedlings (the most common method) or by the direct planting of seeds. Direct seeding is reserved for remote or inaccessible areas where seedling planting is not cost-effective. A few tree species, such as poplars (*Populus* species) and willows (*Salix* species), are artificially reproduced from cuttings. Most forest planting in North America involves the conifers, especially the pines, spruces, and Douglas fir, because of the prospects of successful establishment and high financial yield. The amount of hardwood planted worldwide has increased from earlier periods, with major gains in tropical hardwoods (*Eucalyptus* species, *Gmelina* species) and high value temperate species.

Artificial regeneration offers greater opportunity than natural regeneration to modify the genetic constitution of stands. The most important decision made in artificial regeneration is the selection of the species used in each new stand. The species chosen should be adapted to the site. The most successful introductions are obtained by moving species to the same latitude and position on the continent that they occupied in their native habitat. For example, many conifers of the western coasts of North America have been successful at the same latitudes in western Europe. The forest economy of many countries in the Southern Hemisphere is dependent on pines introduced from localities of comparable climate in the southern United States, California, and Mexico.

The variability of seed quantity and quality and the demand for superior genotypes has led to the creation of seed orchards, stands of trees selected for superior genetic characteristics, which are cultivated to produce large quantities of seed. Most kinds of seed can be stored in sealed containers in refrigerators at temperatures near freezing for several years without a significant loss in viability. For some species, a brief period of cold storage may be necessary for the seeds to germinate; this stratification treatment is needed to satisfy the dormancy requirement of some temperate-zone species.

Direct sowing of harvested seed in the forest or on open land is not a common practice because of forest seed-eaters (mice, squirrels, birds) and the problem of weed growth. Tree seedlings are therefore raised in forest nurseries, where effective protection is possible. These seedlings almost invariably come from seed, although vegetative propagation from rooted cuttings is a useful technique of perpetuating valuable strains of certain species. Seedlings grown in raised seedbeds are removed from the nursery soil when large enough and are bare-rooted when planted in the field. Seedlings grown in individual containers have an intact root system encapsulated in a soil plug for planting. In either case, the system can be highly mechanized. To enhance seedling quality, the seedbeds or container media are inoculated with specific microorganisms that form symbiotic relationships with the seedlings. These microorganisms include certain fungi, which form mycorrhizae with the roots and improve nutrient and water uptake, and nitrogen-fixing organisms such as *Rhizobium* species and *Frankia* species, which contribute nutrients. Selective herbicides, insecticides, and fungicides are applied before or after seedling emergence to keep the developing seedlings free of weeds, insects, and disease.

Many tree seedlings are suitable for field planting after a few months in a containerized seedling nursery or after one to two years in a seedbed. Slow-growing species are transplanted by hand or machine during the dormant season to transplant beds where they are root-pruned and fertilized to stimulate top growth and the development of a bushy root system, characteristics essential for survival in field planting. The mechanized operation is highly efficient. One machine and four workers can transplant 30,000 to 40,000 seedlings each working day. Weeds are controlled during the transplant stage by chemical herbicides that inhibit weed seed germination or growth or by mechanical harrows drawn between the rows.

In preparation for field planting, dormant nursery-grown seedlings are undercut with a sharpened steel blade and removed from the bed by hand or by a mechanized vibrating lifter and conveyor belt system. Roots of seedlings lifted in autumn are packed with moistened sphagnum moss, and the bundles are stored in refrigerated coolers. Alternatively, seedlings may be placed in a trench, or heeling-in bed, and covered with soil and mulch until spring. At the time of lifting, seedlings should be culled to eliminate those that will not survive after planting—that is, seedlings infested with insects or disease, badly damaged in lifting and handling, having distinctly poor root systems, or falling below minimum size standards. It is imperative that the seedlings be kept cool and the root systems moist in all phases of the lifting, storage, transport, and planting processes.

Container-grown seedlings are culled in a manner similar to the bare-rooted stock and in most cases are shipped in the containers in which they were produced. The container method, which has traditionally been used in the tropics or in locations that are hot and dry, has become the principal method of seedling production in

Canada, Scandinavia, and portions of continental Europe, Japan, and China.

Planting tree seedlings is one of the most costly investments in the production of a forest crop. The success of a whole rotation is often determined by the soundness of decisions made about planting. These decisions concern the selection of the planting stock, the density of the planting, the use of mixed plantings, the season of planting, preparation of the site prior to planting, and even the method of planting. In temperate climates planting is generally conducted from late winter to late spring, but the use of container-grown seedlings extends the planting season into the early summer and includes a period in early autumn.

On level ground, machine planting is preferred over hand planting. A planting machine forms a groove in the soil in which seedlings are placed at specified intervals; a set of blades then cuts into the soil around the planted seedling, and a set of packing wheels firms the soil around it. A planting machine pulled behind a single tractor on prepared level ground can set 8,000 to 10,000 seedlings per day. On steep slopes, broken or rocky ground, or amid tree stumps and tops, planting is done by hand. The planter uses a spade, planting bar, or mattock (or a variation of one of these) to cut a notch, or dig a pit, into which the seedling roots are inserted. Soil is then replaced and stamped firmly around the base of the seedling.

During the following growing season, and possibly two to three years thereafter, weed control may be essential for the survival and early growth of the planted seedling. Weeds may be removed by hand with a sharp tool or hoe or by other mechanical means such as mowing or cultivating between the planted rows. Herbicides may offer a more effective and efficient means of weed control. While care must be exercised to shield the tree from many chemicals, compounds are available that kill unwanted vegetation but do not harm the tree seedling. In some regions the lower branches of conifers and certain highly valued hardwoods are pruned from saplings and young trees to improve the quality and value of the main stem and improve access into the plantation. Otherwise, the artificially established plantation needs, and receives, no more attention than does the naturally regenerated crop.

Until the 20th century foresters usually accepted the land much as they found it. Their reaction to infertile soil was to plant aggressive species of trees, regardless of their potential market value, and to accept lower returns in plant production. Development of modern machines and a growing understanding of plant nutrition and soil chemistry now enable foresters to improve sites much as a farmer does and thereby to increase output substantially. Mechanical draining, using tractor-drawn plows to create deep open drains and so aerate the soil, is now usual on the peaty swamps of Europe, especially in Finland. On the hard heathlands of Great Britain, 120,000 hectares (296,526 acres) of new afforestation land were broken up after 1940 with sturdy plows designed to turn over firmly compacted soil layers. Plowing facilitates penetration of air, water, and tree roots, checks weed growth, and lessens fire hazard. So far it has usually been confined to strips for each row of trees, but full plowing as done on a farm promises further advantages.

In the poorly drained Great Lakes states and in coastal areas in the southeastern and southern United States, sites are prepared by a bedding plow, which creates an alternative ridge and valley surface that improves soil drainage, aeration, and nutrient availability. Subsequent to bedding, seedlings are planted on the ridge or bed. Because forest crops are rarely irrigated (returns are too low for the capital cost invested), forest plantings on droughty sites require a careful selection of the species and the time for planting and an effective weed control program.

FORESTS AND GRASSLANDS

A worker plants mangrove trees in a mangrove swamp on the island of Bali in Indonesia. Mangrove swamps can help Indonesia's coastal communities fend off rising seas and stronger tropical storms caused by climate change, experts say. Bloomberg via Getty Images

The fundamental relationship between mineral nutrition and growth is the same for trees as for other plants. An understanding of forest tree nutrition requires recognition of factors distinctive to forests: (1) The nutrient demands of the plantation vary from season to season and with the developmental stage of the stand. During the life of a forest tree crop, large quantities of nutrients are returned to the soil in organic matter, which is, in turn, mineralized and made available for reuse by the same or the following crop. (2) Retranslocation of absorbed nutrients is highly developed in trees; that is, nutrients in leaves move back into stems prior to fall leaf drop and then move into new leaves in the spring. (3) Except for the first year after planting, trees start the growing season with a developed framework for photosynthesis and an established root system for nutrient and water uptake. (4) The use of soil resources such as water and nutrients by trees may often be strongly influenced by mechanisms involved in adaptations for survival from one season to another, rather than in growth.

Judicious management of nutrition ensures not only increased productivity of existing forests but also sustained productivity over many rotations. In southern Australia, for example, declines in yield of 25 to 30 percent in second rotation radiata pine (*Pinus radiata*) plantations have been corrected by a number of means, including intensive silviculture (site preparation, weed control, fertilization) during the early stages, retention and management of forest debris (leaves, branches, etc.) to conserve nutrients, and intercropping with annual legumes, which supply nitrogen and other nutrients.

Range and Forage

Important among the broad spectrum of forest resources are the understory plants that can provide forage for grazing animals, both domestic and wild. Grazing livestock are useful to the forest manager. Dense old-growth forest or vigorous second-growth stands with closed canopies generally have sparse, low-quality forage. Large forest management units, however, generally contain extensive logged or burned areas where understory forage plants temporarily dominate the site. These areas are transitory since the tree canopies close in 10 to 20 years, but they can provide good forage until canopy closure. Cutting cycles in the managed forest and even wildfires provide a continuing grazing resource that shifts from one location to another. In addition, open meadows occurring in valley bottoms, open forests on shallow soils, and grassland balds on windswept ridge tops greatly enrich the grazing potential of the forest. Grazing fees offset the long-term investments that must be carried in renewing the forest.

Hardwood forests are more susceptible than coniferous forests to grazing damage. The current year's growth on broad-leaved trees provides palatable forage during most seasons of the year, whereas coniferous needles are much less palatable. Uncontrolled livestock-grazing in some parts of the world has been particularly devastating to forests and is a serious problem.

Recreation and Wildlife

From the earliest times human beings have looked to the forests for recreation. Today, recreation in forests assumes ever-growing importance with the growth of cities whose inhabitants need a change of scene, fresh air, and freedom to wander, as a relief to the stresses of industrial and commercial life. Imaginative planning is essential to ensure that people actually find what they are seeking without damage to the forest environment or conflict with the pleasures of others. The most popular outdoor recreation activities utilize forestland and include hunting and fishing, picnicking and camping, hiking, mountain climbing, driving for pleasure, boating and other water sports, winter sports, photography, and nature study. The challenge is to balance the varied demands for recreational use with the other forest uses.

For many recreationists the main attraction of the woods is the abundance of animal and plant life. The forest manager must attempt to satisfy the diverse needs of hunters and sportsmen, outdoorsmen, and preservationists. This requires a broad expertise drawing on principles from the social sciences, natural history, wildlife management, landscape design, law, and public administration, among other disciplines.

Recreation management includes visitor management as well as resource management. Reasonably accurate assessments of the type and amount of use that areas receive are important to allow for efficient allocation of budgets and employee time and to ensure that the degree of use does not cause excessive impacts on resources and thus destroy the recreational value of the site. Skillful location of roads, picnic points, parking lots, and campgrounds ensures that the great majority of visitors congregate in relatively small portions of a large forest. Visitor management for some situations can be aided by use of computer-generated simulation models.

Some types of recreation require intensive management and special amenities. Vehicular camping facilities, for example, are designed for intensive use by large numbers of people and typically provide electrical hookups, toilets, showers, picnic tables, fireplaces, garbage receptacles, directional and interpretive signs, and play areas. These features must be durable and easily maintained. Downhill ski areas are most popular when well equipped with various runs, lifts, restaurants, and lodges. Other types of recreation, such as trail hiking and cross-country skiing, demand larger tracts of land but fewer improvements. Wilderness areas afford the personal challenge and serenity of backpacking, tent camping, and canoeing.

The interpretation of what visitors see in the forests has become a growing activity of most forest services. Nature trails, guidebooks, signposts, interpretive museums, and information stations assist visitors who come to learn as well as to enjoy.

Forests contain natural habitats for a wide range of wildlife, from the elks, wolves, lynxes, and bears of northern coniferous forests to the antelopes, giraffes, elephants, lions, and tigers of tropical savannas and jungles. Certain birds, such as pheasants, wood grouse, and quail, have high sporting value, while others are cherished for attractive song, appearance, or rarity. Many endangered species depend on forest habitats that are carefully protected by national and international laws.

Forest managers must attend to the interrelated, and sometimes directly opposed, wildlife interests of hunters, conservationists, and farmers. Obviously the same animal can present a different aspect to each group. A Bengal tiger, for example, provides a biologist with a classic example of a carnivorous beast living in harmony with a jungle environment and restraining its main prey, deer, from undue increase in numbers. But to a village peasant it is a menace to his cows and goats and a threat to the safety of himself and his family, while a game hunter regards it as a magnificent quarry demanding all his skill. The needs of the forest itself require the numbers of grazing and browsing animals to be kept to a tolerable level. Otherwise renewal of tree crops becomes impossible.

Virtually every change that occurs in a forest benefits some wildlife species and harms others. Some species require a diversity of conditions; one type for feeding, another for nesting, and yet another for cover. Some have very specific requirements essential to their existence, whereas others have a broad range of tolerance. In any case, the life history characteristics of the species must be known in order for the resource manager to plan and implement practices necessary for the well-being of the species. Sometimes the best management involves increasing the forest edge habitat, frequented by many kinds of wildlife. Forest edge improvement may be integrated with timber harvesting and the construction of fire lanes and logging roads. Because food and cover for wildlife are often more plentiful in the early stages of forest development, retardation of succession by prescribed burning may be beneficial to wildlife. Food crops may be planted in certain areas to improve the wildlife-carrying capacity. Adjustments are often made by foresters in cutting procedures, rotation age, regeneration methods, and other practices to accommodate the food and cover needs of wildlife and fish. Certain areas may be managed exclusively for wildlife, particularly in situations where habitat for endangered species must be protected.

In virtually every country the sporting aspect of woodland wildlife management is controlled, to some degree, by general game laws, which also apply outside the forests. These prescribe licenses for firearms and the taking of specified birds and beasts; they usually lay down closed seasons during which certain game may not be shot and also set limits to the sportsman's bag of rare species. In the United States a peculiar situation exists whereby the game legislation of the separate states applies unchanged over most publicly owned forests. In other countries the forest managers are in a stronger position, since local game laws are adjusted to their particular requirements.

Watershed Management and Erosion Control

Not only is the presence of water in soils essential to the growth of forests, but improved water yield and quality are becoming increasingly important management objectives on many forested lands. Forests and their associated soils and litter layers are excellent filters as well as sponges, and water that passes through this system is relatively pure. Forest disturbances of various kinds can speed up the movement of water from the system and, in effect, reduce the filtering action. While disturbances are inevitable, in most instances they need not contribute to poor water quality.

In mountainous territory the value of forests for watershed and erosion protection commonly exceeds their value as sources of lumber or places of recreation. The classic example is found in Switzerland and the neighbouring Alpine regions where the existence of pastoral settlements in the valley is wholly dependent on the maintenance of continuous forest cover on the foothills of the great peaks. This is combined skillfully with limited lumbering and widespread recreational use by tourists.

The guiding principle of management where erosion threatens is therefore the maintenance of continual cover. Ideally, this is achieved by single-stem harvesting; only one tree is felled at any one point, and the small gap so created is soon closed by the outward growth of its neighbours.

The progress of water, from the time of precipitation until it is returned to the atmosphere and is again ready to be precipitated, is called the hydrologic cycle. The properties of the soil plant system provide mechanisms that regulate interception, flow, and storage of water in the cycle. The water that moves downward into the soil, or infiltrates, is the difference between precipitation and the losses due to canopy interception, forest floor interception, and runoff. The amount of water stored in the soil is largely dependent on the physical properties of the soil, its depth, and the amount of water lost due to evaporation from the soil surface and transpiration from plants (evapotranspiration). Transpiration is the water absorbed by plant roots that is subsequently evaporated from their leaf surfaces. Deep forest soils have a high water-storage capacity. Unless they are very porous and drain freely, they have a water table below which the subsoil is saturated. The depth of the water table varies seasonally and is higher during periods of low evapotranspiration. Removal of the forest canopy in wet areas also raises the water table. Most tree roots need air to survive and cannot exploit soil below the water table. The drainage of land having a high water table usually increases the productivity of the forest.

When incoming precipitation exceeds the soil's water-storage capacity, the excess water flows from the soil and can be measured as streamflow. The water yield of a forest is a measure of the balance between incoming precipitation and outflow of water as streamflow. The amount of increase in water yield depends on annual precipitation as well as the type and amount of overstory vegetation removed. As forests regrow following cutting, increases in streamflow decline as a result of increased transpirational losses. Streamflow declines are greater in areas that are restocked with conifers than in those restocked with hardwoods. This results from greater transpiration losses during the winter months from coniferous species.

Despite the uncertain balance of water gain and loss, forests offer the most desirable cover for water management strategies. Water yields are gradual, reliable, and uniform, as contrasted to the rapid flows of short duration characteristic of sparsely vegetated land. Unforested land sheds water swiftly, causing sudden rises in the rivers below.

Over a large river system, such as that of the Mississippi, forests are a definite advantage since they lessen the risk of floods. They also provide conditions more favourable to fishing and navigation than does unforested land. All natural streams contain varying amounts of dissolved and suspended matter, although streams issuing from undisturbed watersheds are ordinarily of high quality. Waters from forested areas are not only low in foreign substances, but they also are relatively high in oxygen and low in temperature. Nonetheless, some deterioration of stream quality can be noted during and immediately after clear-cut harvesting, even under the best logging conditions. The potential for water-quality degradation following timber harvest may involve turbidity (suspended solids) as well as increases in temperature and nutrient content. Sediment arising from logging roads is the major water-quality problem related to forest activities in many areas.

The belief that forests increase rainfall has not been substantiated by scientific inquiry. Local effects can, however, prove substantial, particularly in semiarid regions where every millimetre of rain counts. The air above a forest, as contrasted with grassland, remains relatively cool and humid on hot days, so that showers are more frequent. Fog belts, such as those found along the Pacific seaboard of North America and around the peaks of the Canary Islands, give significant water yields through the interception of water vapour by tree foliage. The vapour condenses and falls in a process described as fog drip.

Fire Prevention and Control

A forest fire is an unenclosed and freely spreading combustion that consumes the natural fuels of a forest; that is, duff, grass, weeds, brush, and trees. Forest fires occur in three principal forms, the distinctions depending

FOREST FIRES

A forest fire is an uncontrolled fire occurring in vegetation more than 1.8 metres (6 feet) in height. These fires often reach the proportions of a major conflagration and are sometimes begun by combustion and heat from surface and ground fires. A big forest fire may crown—that is, spread rapidly through the topmost branches of the trees before involving undergrowth or the forest floor. As a result, violent blowups are common in forest fires, and they may assume the characteristics of a firestorm.

The Zaca wildfire in Santa Barbara County, southwestern California, 2007. John Newman/U.S. Forest Service

essentially on their mode of spread and their position in relation to the ground surface. Surface fires burn surface litter, other loose debris of the forest floor, and small vegetation; a surface fire may, and often does, burn taller vegetation and tree crowns as it progresses. Crown fires advance through the tops of trees or shrubs more or less independently of the surface fire and are the fastest spreading of all forest fires. Ground fires consume the organic material beneath the surface litter of the forest floor; ground fires are the least spectacular and the slowest-moving, but they are often the most destructive of all forest fires and also the most difficult to control.

A forest fire does a number of specific things. First, and perhaps most obviously, it consumes woody material. Second, the heat it creates may kill vegetation and animal life. In most fires, much more is killed, injured, or changed through heat than is consumed by fire. Third, it produces residual mineral products that may cause chemical effects, mostly in relation to the soil. The lethal temperatures for the living tissues of a tree (that is, the phloem and cambium, which are located under the bark) begin at 49 °C (120 °F) if exposure is prolonged for one hour. At 64 °C (147 °F) death is almost instantaneous. The ignition temperature for woody material is approximately 343 °C (650 °F), with a flame temperature of 870 to 980 °C (1600 to 1800 °F).

Forest fires seldom occur in tropical rainforests or in the deciduous broad-leaved forests of the temperate zones. But all coniferous forests, and the evergreen broad-leaf trees of hot, dry zones, frequently develop conditions ideally suited to the spread of fire through standing trees. For this, both the air and the fuel must be dry, and the fuel must form an open matrix through which air, smoke, and the gases arising from combustion can quickly pass. Hot, sunny days with low air humidity and steady or strong breezes favour rapid fire spread. In coniferous forests the resinous needles, both living and dead, and fallen branch wood make an ideal fuel bed. The leaves of evergreen broadleaf trees, such as hollies, madrone, evergreen oaks, and eucalyptus, are coated in inflammable wax and blaze fiercely even when green. Once started, fire may travel at speeds of up to 15 km (10 miles) per hour downwind, spreading slowly outward in other directions, until the weather changes or the fuel runs out.

Well over 95 percent of all forest fires are caused by people, while lightning strikes are responsible for 1 to 2 percent. In some countries the setting of fires for clearing cropland is an integral technique of agriculture. In other areas forest fire prevention, including public education, hazard reduction, and law enforcement, consumes a considerable amount of time and money. The two basic steps in preventing forest fires are reducing risk and reducing hazard. Risk is the chance of a fire's starting as determined by the presence of activity of causal agents, most likely human beings. Hazard is reduced by compartmentalizing a forest with firebreaks (alleyways in which all vegetation is removed) and reducing the buildup of fuel (litter, branches, fallen trees, etc.) by controlled burning. In the United States the Forest Service devised a National Fire-Danger Rating System, which is the resultant of both constant and variable fire danger factors that affect the inception, spread, and difficulty of control of fires and the damage they cause.

Effective fire control begins with a field survey and map to identify the areas at risk, delineate them, and define and improve the barriers or firebreaks that may limit fire spread. Natural barriers include rivers, lakes, ridge tops, and tracts of bare land. Artificial barriers can be roads, railways, canals, and power-line tracks, but usually extra firebreaks must be cut to link these and provide wider gaps that fire cannot readily jump. Belts of land from 10 to 20 metres wide are cut clear of trees or left unplanted when a new forest is formed. Sometimes the soil is left bare and cultivated only at intervals to check invasion by weeds. Usually it is sown with an even crop of low perennial grasses or clovers and kept short by mowing or grazing. This checks soil erosion, provides an evergreen fireproof surface, and allows access on foot, by car, or in an emergency by fire-fighting trucks. Surfaced roads, serving also for lumber haulage and access for recreation, are of critical importance in fire fighting. Signposts are needed to guide fire crews unfamiliar with the woods and to mark water supplies and rendezvous points.

Detection is the first step in fire suppression. Many countries have organizations of trained professionals to detect and fight fires; others rely on volunteers or a combination of the two. Tower lookouts are the mainstay of nearly all detection systems, although the use of aircraft and satellites has modified this view in countries with an advanced fire control program. Fire surveillance is essential during seasons of high risk. Towers are set on hilltops where observers equipped with binoculars, maps, and a direction scale determine the compass direction of smoke and notify the fire control base via telephone or radio. If a fire can be seen from two or more towers, its precise position is quickly determined by mapping the intersection of cross bearings. Aircraft are used to detect fires and to carry out reconnaissance of known fires. Aerial surveillance has probably been most successful in detecting lightning-caused fires and is most often employed in areas of relatively low-value lands and inaccessible areas. An aircraft is essentially a moving fire tower, and the problems of detection that apply to a tower also apply to an aircraft; however, new developments in remote-control television, high-resolution photography, heat-sensing devices, film, and radar make fire detection by aircraft and satellite more efficient and location more accurate. Satellites provide a rapid means of collecting and communicating highly precise information in fire detection, location, and appraisal.

Once a fire has been detected, the next step is fire suppression. The first job is to stop or slow the rate of spread of the fire, and the second job is to put it out. The aim of suppression is to minimize damage at a reasonable cost. This does not necessarily mean the same thing as minimizing the area burned, but it is a major goal. Suppression is accomplished by breaking the "fire triangle" of fuel, temperature, and oxygen by robbing the fire of its fuel (by physically removing the combustible material or by making it less flammable through application of dirt, water, or chemicals); by reducing its temperature (through application of dirt, water, or chemicals and partial removal or separation of fuels); and by reducing the available oxygen (by smothering fuels with dirt, water, fog, or chemical substances).

The great majority of all forest fires are contained by professional fire fighters equipped with numerous hand

A firefighter hikes up a hill after lighting a backfire in an attempt to stop a larger fire from burning out of control in in Arrowbear Lake, California, in 2007. Justin Sullivan/Getty Images

tools (spades, beaters, axes, rakes, power saws, and backpack water pumps). Trained fire crews with light hand equipment can be carried quickly to a fire by truck, delivered by helicopter, or even dropped by parachute. When necessary, large machines (bulldozers or plows) are used to clear openings, or firebreaks, which stop the spread of the fire. This requires clearing surface and sometimes aerial fuels from a strip of land and then digging down to mineral soil to stop a creeping or surface fire. A control line can also be established by directly extinguishing the fire along the edge or by making fuels nonflammable. In some cases a backfire may be deliberately set between the control line and the oncoming fire to burn out or reduce the fuel supply before the main fire, or head fire, reaches the control line.

Water is the most obvious, efficient, and universal fire extinguisher, but large-scale use of water in fire fighting is limited because it is usually in short supply and application methods are not adequate. For these reasons other materials have been tested for persistence and efficiency in putting out fires. Wetting agents change the physical characteristics of water to increase its penetrating and spreading abilities. Retardants, such as sodium calcium borate, reduce the flammability of wood and therefore its rate of burning. Foaming agents in powder or liquid form can greatly increase the mixture volume and thereby cool, moisten, and insulate the fuel.

Aircraft can quickly carry in water and other chemicals to be dropped or sprayed on the fire. A method developed on Canadian lakes is to fill the floats of a seaplane with water, which is done as it skims the lake on takeoff, and to discharge this through nozzles over the fire.

The prescribed use of fire in forestland management is approached with understandable reluctance by many foresters and wildland managers. Yet, fire has a place in the management of particular ecosystems. The decision to use fire is usually based on a balancing of pros and cons; that is, damage, possible or expected, must be weighed against benefits. Under proper circumstances, prescribed burning can be used to prepare seedbeds for natural germination of most tree species, to control insect and disease infestations, to reduce weed competition, to reduce fire hazard, and to manipulate forest cover type.

Insect and Disease Control

Enormous numbers and varieties of insects, fungi, bacteria, and viruses occur in forests and are adapted to live on or around trees. Many of these are beneficial, and even the destructive ones are usually held in check by their natural enemies or an unfavourable environment. The normal population levels of pest organisms result in limited reduction in tree growth or the total destruction of only a small number of trees in the forest. The losses are generally accepted by foresters as unavoidable and are tolerated as long as the annual destruction does not seriously affect the net annual increase in wood production.

Every part of a growing tree—root, trunk, bark, leaf, flowers, and seeds—is potentially subject throughout every stage of its life to attack by some harmful insect or fungus. Insects actually destroy more standing timber than does any other agent. Bark beetles, including species of *Dendroctonus* and *Ips*, are among the most destructive insects. They bore into the tree and feed just below the bark, where they create tiny channels that disrupt the flow of food to the roots, often killing the tree. Diseases frequently retard growth of trees and are less of a factor in mortality. A particularly destructive disease is caused by fungi that decay the wood of trees. The heart-rot fungi gain entrance through any wound resulting from fire scars, broken limbs, or anything else that damages the tree's protective tissue. Were it not for heartrot, a large number of conifers and broad-leaved trees could be left to grow for many more years.

Insect and disease organisms accidentally introduced to forests from other parts of the world often develop serious epidemic conditions because of the lack of any natural control. Because of rapid global transportation, insects and fungal spores can be spread easily throughout the world and arrive in a healthy condition. The seriousness of the situation cannot be overestimated, and the enforcement and improvement of plant quarantine laws is essential. Typically disasters have arisen where quarantine has failed or has been imposed too late. The American chestnut, *Castanea dentata*, has been virtually wiped out by the chestnut blight fungus, *Endothia parasitica*, which does little harm to related trees in its native China. Elms have suffered severely, both in Europe and in the United States, from the elm disease fungus, *Ceratocystis ulmi*, which was first detected in the Netherlands and is carried from tree to tree by flying beetles. Minute aphids, probably introduced on living plants from Asia, now make it impossible to raise commercial crops of two conifers once valued in Britain, namely, the white pine, *Pinus strobus*, from New England, and the European silver fir, *Abies alba*, native to Switzerland.

Generally the healthier the forest, the more resistant it is to widespread pest attack. Overmature, weak, windthrown, and lightning- or fire-killed trees have little or no defense against infestation and are a factor in the buildup of pest populations. Selective cutting of susceptible trees, thinning that accelerates growth, and other similar long-range forest management practices that stimulate vigorous tree growth are good methods for indirect control of insects and diseases. These practices reduce the host material and breeding grounds of pests that may spread to healthy trees. In regions with a high incidence of a known pest, foresters attempt to avoid serious trouble by planting only trees known to resist existing pests in the regions where the trees are grown. Many forest genetic programs have as a major goal the selection and breeding of trees with insect and disease resistances.

Occasionally the natural conditions that suppress the population of pest organisms change, and outbreaks in forests may reach epidemic proportions. Even-aged stands and plantations with trees of the same species and of uniform size and age often create perfect conditions for the rapid spread of insects and diseases. Even the more complex uneven-aged forests with their inherent check-and-balance systems can develop devastating populations of pests. At this point the forester must consider direct control measures.

Because effective direct control of insects and diseases of standing timber is generally expensive, it is employed only when the potential mortality or loss in growth is extreme. Routine monitoring of insects and diseases allows foresters to schedule timely harvests of infested trees and to limit the spread of the problem to uninfested trees or areas. These sanitation and salvage harvests, coupled with piling and burning the limbs and branches left after logging, reduce the material and conditions that allow pest populations to develop. Traps baited with sex-attractant chemicals, or pheromones, are a promising method to reduce breeding populations of certain insects. Application of insecticidal or fungicidal sprays from the ground or from low-flying aircraft offer a short-term measure to check sudden plagues of insects or outbreaks of fungal diseases. Action has most frequently been taken against exceptional outbreaks of defoliating caterpillars, including those of the gypsy moth in the United States, the nun moth in central Europe, and the pine looper moth in England. At the time of year when feeding caterpillars are most vulnerable, light aircraft fly across the forest on carefully planned courses, distributing pesticides.

A disadvantage of these blanket treatments by potent, broad-spectrum chemicals is that they also eliminate parasitic and predatory insects that serve as natural controls on the pest's numbers; they may also adversely affect birdlife. In practice, large-scale chemical treatments of forests are infrequent and are restricted to a small proportion of the areas at risk. Generally, natural control through predatory organisms, which also cycle opportunistically in a slightly delayed sequence with the pest populations, combined with physical factors like cold winters, provides adequate checks. Biological control involving the release of predators or diseases of pests is promising in some situations.

Less spectacular preventive measures are commonly taken as routine steps in practical forestry to lessen anticipated losses. Nursery stock, easily reached and handled, may be grown in fumigated seedbeds and sprayed during production so that uninfested and vigorous seedlings are planted in the forests. Prompt removal of logs from the forest to distant processing mills transfers beetles that may emerge from beneath the bark to areas where they can do no harm. Stumps of freshly felled conifers can be easily and cheaply treated by brushing on a fungicide to check the white root-rot fungus, *Fomes annosus*. This is a serious agent of decay that spreads underground through root grafts after gaining entry via the exposed surface of a felled stump.

Agroforestry

Agroforestry is a practice that has been utilized for many years, particularly in developing countries, and is now widely promoted as a land-use approach that yields both wood products and crops. Trees and crops may be grown together on the same tract of land in various patterns and cycles. The trees may be planted around the perimeter of a small farm to provide fuelwood and to serve as a windbreak. The limbs and foliage may be removed periodically for livestock fodder. Trees also may be planted in rows that alternate with crops or they may be planted more densely with interplanting of crops until crown closure of the trees precludes further crop production. These practices are most extensively used as a part of subsistence agriculture, but their use in large-scale production systems is becoming more common.

Urban Forestry

Urban forestry, which is the management of publicly and privately owned trees in and adjacent to urban areas, has emerged as an important branch of forestry. Urban forests include many different environments such as city greenbelts; street and utility rights-of-way; forested watersheds of municipal reservoirs; and residential, commercial, and industrial property. An important distinction between urban and rural forestry is that urban trees are more highly valued than rural trees and often receive expensive individual care and attention. Many professional foresters are trained to handle the special problems of urban trees and to foster the diverse benefits they provide.

CHAPTER 3

Grasslands

Grasslands are areas in which the vegetation is dominated by a nearly continuous cover of grasses. Grasslands occur in environments conducive to the growth of this plant cover but not to that of taller plants, particularly trees and shrubs. The factors preventing establishment of such taller, woody vegetation are varied.

Grasslands are one of the most widespread of all the major vegetation types of the world, covering nearly 40 percent of Earth's land surface. This is so, however, only because human manipulation of the land has significantly altered the natural vegetation, creating artificial grasslands of cereal crops, pastures, and other areas that require some form of repetitious, unnatural disturbance such as cultivation, heavy grazing, burning, or mowing to persist. This discussion, however, concentrates on natural and nearly natural grasslands.

Although not usually associated with grasslands, tundras are also included in this chapter. Tundra ecosystems are similar to grasslands in that they are treeless and made up of low vegetation. Unlike grasslands, tundras are composed of low shrubs, lichens, and mosses and occur in high-latitude and high-altitude terrestrial environments.

THE ORIGIN OF GRASSLANDS

The most extensive natural grasslands can be thought of as intermediates in an environmental gradient, with forests at one end and deserts at the other. Forests occupy the most favourable environments, where moisture is adequate for growth and survival of a tall, dense vegetation dominated by trees. Deserts are found where moisture is so lacking that a continuous, permanent vegetation cover cannot be maintained. Grasslands lie between these two extremes.

Like the savannas, deserts, and scrublands into which they commonly blend, grasslands arose during the period of cooling and drying of the global climate, which occurred during the Cenozoic Era (65.5 million years ago to the present). Indeed, the grass family itself (Poaceae or Gramineae) evolved only early in this era. The date of earliest appearance of grasslands varies from region to region. In several regions a succession of vegetation types can be recognized in the Cenozoic fossil record, as climate dried out progressively. For example, in central Australia during the past 50 million years tropical rainforest gave way successively to savanna, grassland, and, finally, desert. In some places expansion of grasslands to something approaching their modern

Richardson's masonhalea lichen (Masonhalea richardsonii). *This lichen can be found in Alaska's tundra.* Ben Strickland—Van Cleve Photography

extent occurred only during the extremely cold, dry intervals—called ice ages in north temperate regions—of the past two million years.

A dynamic balance commonly exists between grasslands and related vegetation types. Droughts, fires, or episodes of heavy grazing favour grassland at some times, and wet seasons and an absence of significant disturbances favour woody vegetation at others. Changes in the severity or frequency of these factors can cause a change from one vegetation type to another.

Other grassland types occur in places too cold for trees to grow—that is, beyond the forest limits of high mountains or at high latitudes. A characteristic type of grassland in cool, moist parts of the Southern Hemisphere is tussock grassland, dominated by tussock or bunch grasses that develop pedestals of matted stems, giving the vegetation a lumpy appearance. Tussock grasslands occur at various latitudes. In the tropics they are found above the forest limit on some high mountains—such as in New Guinea and East Africa. At the higher latitudes of the Southern Ocean they form the main vegetation of subantarctic islands. They are also typical of the drier, colder parts of New Zealand and the southernmost regions of South America.

Not all natural grasslands, however, arise from climate-related circumstances. Woody plants may be prevented from growing in certain areas for other reasons, allowing grasses to dominate. One cause is seasonal flooding or waterlogging, which is responsible for the creation and maintenance of large grasslands in parts of the highly seasonal subtropics and in smaller areas of other regions. One of the best examples of a seasonally flooded subtropical grassland is the Pantanal in the Mato Grosso region of Brazil. Across an area of 140,000 square km (54,000 square miles), dry grasslands prevail for half of each year and shallow wetlands for the other, with small forest patches restricted to low rises that do not flood during the wet season. In many other areas where climate is suitable for forest growth, very shallow or infertile soils may prevent tree growth and result in development of grassland.

The largest areas of natural grassland—those resulting from climatic dryness—can be classified into two broad categories: tropical grasslands, which generally lie between the belts of tropical forest and desert; and temperate grasslands, which generally lie between deserts and temperate forests. Tropical grasslands occur in the same regions as savannas, and the distinction between these two vegetation types is rather arbitrary, depending on whether there are few or many trees. Likewise, temperate grasslands may have a scattering of shrubs or trees that blurs their boundaries when they occur adjacent to scrublands or temperate forests.

Tropical grasslands are found mainly in the Sahel south of the Sahara, in East Africa, and in Australia. Temperate grasslands principally occur in North America, Argentina, and across a broad band from Ukraine to China, but in most of these regions they have been substantially altered by agricultural activities.

Many grasslands formerly supposed to be natural are now recognized as having once been forests that grew in a marginally dry climate. Early human disturbance is responsible for their transformation. For example, almost the entire extensive lowland grasslands of the eastern part of the South Island, New Zealand, are believed to have been created by forest burning carried out by the Polynesians— the country's first colonists—during the eight centuries before European settlement in the 18th century.

Seminatural grasslands may occur where woody vegetation was once cleared for agricultural purposes that have since been abandoned; a return to the original vegetation is prevented by repeated burning or grazing. In wet tropical regions these types of grasslands may be very dense, such as those in East Africa that are dominated by elephant grass (*Pennisetum purpureum*) or in New Guinea by pit-pit grass (*Miscanthus floridulus*), both of which grow 3 metres (10 feet) tall.

All areas of grassland may owe something of their area and character to a long history of interaction with humans, particularly through the medium of fire.

GRASSLAND CLIMATES AND SOILS

Grassland climates are varied, but all large regions of natural grassland are generally hot, at least in summer, and dry, though not to the extent that deserts are. In general, tropical grasslands receive 500 to 1,500 mm (20 to 60 inches) of rain in an average year and in every season experience temperatures of about 15 to 35 °C (59 to 95 °F). The dry season may last as long as eight months. An excess of rainfall over evaporation, leading to ephemeral river flow, occurs only during the wet season. The tropical grassland climate overlaps very broadly with that of savanna. As previously stated, these vegetation types differ little from each other, a savanna being merely a grassland with scattered trees. Small changes in management and usage can convert one to the other.

Temperate grasslands are somewhat drier than tropical grasslands and also colder, at least for part of the year. Seasonal temperature variation may be slight in tropical grasslands but may vary by as much as 40 °C (72 °F) in temperate grassland areas. Mean annual rainfall in the North American grassland areas is 300 to 600 mm (12 to 24 inches). Mean temperatures in January range from -18 °C (0 °F) in the north to 10 °C (50 °F) in the south, with corresponding values in July being 18 °C (64 °F) and 28 °C (82 °F). Mean annual temperature in the most northerly areas of the North American grassland zone is below 0 °C (32 °F).

Occurring as they do across a wide range of climatic and geologic conditions, grasslands are associated with many different types of soil. The grassland ecosystem itself influences soil formation, and this causes grassland soils to differ from other soils. The nature of grass litter and its pattern of decomposition commonly result in the development of a dark, organically rich upper soil layer that can reach 300 mm (12 inches) below the surface. This layer is absent from desert soils and is different from the surface layer of rotting leaf litter typical of forest soils. It is friable in structure and rich in plant nutrients. Lower soil layers are typically pale and yellowish, especially at depths close to two metres.

GRASSLAND BIOTA

In pre-European North America, grasslands spread across a large portion of the continent, from the Rocky Mountains in the west to the deciduous forests in the east. Of this vast expanse, only tiny fragments remain in any condition remotely similar to their original state. The largest central area consisted of mixed prairie, dominated by several species of the grasses *Stipa*, *Agropyron*, *Bouteloua*, and *Koeleria*. Mixed prairie gave way in the north to a fescue prairie with *Festuca* and *Helictotrichon*; in the west, to a short grass steppe dominated by *Bouteloua gracilis* and *Buchloe dactyloides*; and to the east, to a tall grass prairie with the bluestem grasses *Andropogon gerardii* and *A. scoparium*. Trees and shrubs were generally absent, but a large variety of herbaceous plants occurred with the grasses.

The large grazing mammals of the North American prairies included the bison and pronghorn antelope, whose typical predator was the gray wolf. The badger and several rabbit and hare species were widespread, as were many small burrowing rodents. Among the invertebrate fauna, grasshoppers were and still are particularly important. In grasslands the total biomass of invertebrates typically exceeds that of the much larger and more conspicuous vertebrates, except in regions containing large numbers of domestic stock.

The principal region of grassland in South America is in the southeast portion of the continent; it can be divided into the Pampas of Argentina and the campos of the adjacent areas of Uruguay and Brazil. Among the many grasses in the Pampas, *Stipa* is the most diverse, while another suite of grasses in the campos includes among its more common members *Paspalum* and *Andropogon*. The vegetation of the region is now greatly altered by centuries of heavy grazing and burning. Before this, the principal large herbivore was the pampas deer, of which only one small herd survives. Opossums, armadillos, and rodents also were abundant and were preyed upon by various cats, such as jaguars and pumas, and foxes. The rhea, a large, flightless bird, is indigenous to this area.

The Eurasian steppe, occupying an extensive tract of the former Soviet Union and Mongolia, is in many ways very similar to the prairie of North America and hence harbours many similar plants and animals. Various species of *Stipa* dominate the flora in most areas, and they are mixed in spots with other grasses, of which *Festuca* and *Agropyron* assume dominance. Through their feeding and burrowing activities rodents are important to the maintenance and composition of the vegetation; species include the large marmots and a diversity of voles and other smaller types. A vole in Mongolia, *Lasiopodomys brandtii*, in some years can consume such a high proportion of the vegetation that it reduces its grassland habitat to virtual desert.

The Sahel—the broad band of grassland crossing western and north-central Africa south of the Sahara—is the largest area of tropical grassland. For millennia human populations have put many demands on the region, so that its present condition is quite unlike its natural condition. The most common grasses include *Aristida*, *Cenchrus*, and *Schoenefeldia*. Other species, which are highly palatable to grazing animals, are now restricted to rocky sites that offer some protection; these species may have once been far more widespread and important. In many places where shrubs and small trees occur the vegetation would be called more accurately a scrubland or savanna were it not so easily transformed into grassland by practices such as grazing, burning, and fuel gathering.

The grasslands of East Africa include wetter environments than exist in the Sahel and hence are more diverse. Where forests have been destroyed, a tall grassland consisting of *Pennisetum* or *Hyparrhenia* develops and may be kept in this condition indefinitely through burning or through the browsing and grazing of such herbivores as elephants. Other grasses such as *Aristida* and *Chrysopogon* are important in drier sites, and *Themeda* occurs in cooler places at higher altitudes. Herbivorous mammals include wildebeests, several antelope species, and—where they still survive—rhinoceroses, buffalo, and elephants. Carnivores include various dogs (jackals), cats (cheetahs, lions), hyenas, and mongooses.

Tropical grasslands in Australia in the extensive arid areas are generally dominated by species of the spinifex grasses, *Plectrachne* and *Triodia*, which form characteristic hummocks by trapping windblown sand at the bases of their tussocks. *Heteropogon* and *Sorghum* dominate grasslands in moister, northern areas, and *Astrebla* (Mitchell grass) is prevalent in seasonally arid areas, especially on cracking clay soils in the east. Other grass species are usually subordinate but may dominate in spots. Woody plants, particularly *Acacia* in arid areas and *Eucalyptus* in moister places, may be so numerous that the vegetation cannot be considered true grassland. The Mitchell grasslands were once much purer until they were altered by heavy grazing of domestic stock; today, vast tracts have been invaded by the African shrub *Acacia nilotica*, introduced by humans.

The largest native animals in Australian grasslands are kangaroos, of which the biggest is the red kangaroo, characteristic of the dry inland areas where natural grasslands are found. Mammals introduced from other continents, however, have become as common; these include domesticated stock, especially cattle and sheep, and a range of feral fauna such as camels, horses, donkeys, and goats. European rabbits are also widespread, abundant, and highly destructive. A main predator is the dingo, or wild dog. Reptiles, especially lizards, are very diverse; birds include the large flightless emu as well as a wide range of parrots and other flying forms.

The tussock grasslands of New Zealand and the subantarctic islands are commonly dominated by species of *Poa*. Related vegetation also occurs on high mountains in equatorial regions. These grasses have in most cases evolved in the absence of grazing mammals; the introduction of exotic fauna such as deer and sheep has caused severe degradation of the vegetation in many places.

GRASSLAND POPULATION AND COMMUNITY DEVELOPMENT AND STRUCTURE

Whether tropical or temperate, natural grasslands occur in environments in which growing conditions are

Vegetation profile of a grassland. Encyclopædia Britannica, Inc.

favourable for only a short season. In tropical regions this growing season is usually the rainy season or, in some cases, the season when the ground is not waterlogged or submerged. In temperate grasslands the growing season is usually the short period between the cold, damp winter and the hot, dry summer. Perennial grasses, relying on subterranean reserves of stored food for rapid shoot growth, are well adapted to exploiting such brief growing seasons, reaching their maximum size and completing their seeding within a few weeks. Their aboveground parts then die back, providing potential fuel for the grass fires that typify these environments. The underground perennating roots and rhizomes of the grasses, however, are relatively well protected from fire.

Grasslands tend to produce larger amounts of new growth if subjected to some type of repeated disturbance, usually grazing or fire, that prevents the accumulation of a thick layer of dead litter. Where such a layer is allowed to develop, it retains nutrients in a form not immediately available to roots and acts as a physical barrier for new shoots growing from the soil surface toward the light; in temperate grasslands this layer acts as thermal insulation, slowing the spring warming of the soil. This has obvious implications for grazing management of these systems.

Woody plants, which might be expected to shade the grasses and dominate the vegetation, are disadvantaged by the shortness of the growing season. Nevertheless, in the absence of heavy mammalian grazing and especially of regular fires, some trees and shrubs that grow vigorously may become established. Thus, the grasslands in such situations are maintained by these natural, or seminatural, disturbances of fire and grazing, which prevent the succession of the grassland vegetation toward tropical deciduous forest, savanna, scrubland, or temperate forest.

THE BIOLOGICAL PRODUCTIVITY OF GRASSLANDS

Because of its importance for grazing and other grassland agricultural production, grassland productivity has been extensively investigated using various methods. However, most studies have focused only on aboveground productivity, ignoring the important subterranean component, which can be much more substantial—as much as 10 times greater—even when the aboveground portion is at a seasonal maximum. Typical aboveground biomass (dry weight of organic matter in an area) values for North American grasslands are 2.5 to 6 metric tons per hectare, of which about three-quarters is in the form of dead shoot material; however, values up to and greater than 20 metric tons per hectare have been recorded in some tropical grasslands.

There is a relatively rapid turnover of plant matter in grasslands. Most plant organs have a life span of only one or a few seasons, leading to annual rates of overall turnover of about 20 to 50 percent. When consumption of plant parts by consumers is taken into account, annual

GRASSES

Grass is the collective name for any of many low, green, nonwoody plants belonging to the grass family (Poaceae), the sedge family (Cyperaceae), and the rush family (Juncaceae). There are many grasslike members of other flowering plant families, but only the approximately 10,000 species in the family Poaceae are true grasses.

They are economically the most important of all flowering plants because of their nutritious grains and soil-forming function, and they also have the most widespread distribution and the largest number of individuals. Grasses provide forage for grazing animals, shelter for wildlife, construction materials, furniture, utensils, and food for humans. Some species are grown as garden ornamentals, cultivated as turf for lawns and recreational areas, or used as cover plants for erosion control. Most grasses have round stems that are hollow between the joints, bladelike leaves, and extensively branching fibrous root systems.

productivity values are similar to biomass values. Aboveground annual productivity in a typical Canadian grassland was found to be 6.4 metric tons per hectare when all leaching and decomposition losses from dead shoots were taken into consideration. Values found in other grasslands have varied greatly but often have been significantly lower than this, partly because such losses have not been fully taken into account. Repeated harvesting (surrogate grazing) commonly yields 1.5 to 2.5 metric tons per hectare per year.

Grasslands frequently have been converted to cropland on which edible grains are grown; this allows food for humans to be taken directly from the grasslands themselves rather than via grazing animals feeding on the native grasses in a rangeland situation. The increase in yields is substantial. According to figures compiled for the northern Great Plains of North America, for example, one hectare of natural grassland grazed by cattle can provide the equivalent of one-tenth of one person's annual food energy requirements, whereas the same area sown to wheat can provide four persons' requirements.

However, it should be pointed out that wheat yields cannot be sustained annually at their initially high levels without additional inputs of fertilizer. Similarly, sustained heavy grazing of natural grasslands can quickly lead to degradation of the vegetation and erosion of the soil. This is particularly likely to occur unnoticed in regions that lack fences and therefore do not exhibit fenceline contrasts in range condition. In these fenceless grazing regions degradation due to overgrazing may be difficult to distinguish from background changes due to long- or short-term weather variations.

PRAIRIES

Prairies are level or rolling grasslands, especially those found in central North America. Decreasing amounts of rainfall, from 100 cm (about 40 inches) at the forested eastern edge to less than 30 cm (about 12 inches) at the desertlike western edge, affect the species composition of the prairie grassland. The vegetation is composed primarily of perennial grasses, with many species of flowering plants of the pea and composite families. Most authorities recognize three basic subtypes of prairie: tallgrass prairie; midgrass, or mixed-grass, prairie; and shortgrass prairie, or shortgrass plains. Coastal prairie, Pacific or California prairie, Palouse prairie, and desert plains grassland are primarily covered with combinations of mixed-grass and shortgrass species.

Tallgrass prairie, sometimes called true prairie, is found in the eastern, more humid region of the prairie that borders deciduous forest. The rich soil is laced with the deep roots of sod-forming tallgrasses such as big bluestem and prairie

Grasslands

*Saltmeadow cordgrass (*Spartina patens*)*. Grant Heilman Photography

cordgrass, or slough grass, in the wet lowlands and the shorter roots of bunchgrasses such as needlegrass, or porcupine grass, and prairie dropseed on the drier upland sites.

Midgrass, or mixed-grass, prairie, supporting both bunchgrasses and sod-forming grasses, is the most extensive prairie subtype and occupies the central part of the prairie region. Species of porcupine grass, grama grass, wheatgrass, and buffalo grass dominate the vegetation. Sand hills are common in the western portion bordering the shortgrass plains.

Shortgrass plains occupy the driest part of the prairie and are covered primarily by species of buffalo grass and grama grass. Kentucky bluegrass, although not a native prairie species, is found in all three major prairie subtypes.

The bison, wolf, and most prairie chickens have disappeared from the prairie; but the coyote, prairie dog, jackrabbit, badger, horned lark, meadowlark, and various species of hawks and waterfowl are still common. Insects also are abundant, especially grasshoppers and flies.

In the past, a combination of high summer temperatures, strong winds, late summer drought, and accumulations of dead vegetation set the stage for many naturally caused fires, which prevented trees from becoming abundant in prairie vegetation. Now the fertile prairie soils (brunizem, chernozem, chestnut, and brown soils) are intensely cultivated (primarily corn in the eastern part and wheat in the central area) or grazed (especially the shortgrass region), and little native prairie remains, other than in small protected patches.

SAVANNAS

Savannas make up a vegetation type that grows under hot, seasonally dry climatic conditions and is characterized by an open tree canopy (that is, scattered trees) above a continuous tall grass understory. The largest areas of savanna are found in Africa, South America, Australia, India, the Myanmar-Thailand region, and Madagascar.

The Origin of Savannas

Savannas arose as rainfall progressively lessened in the peripheral regions of the tropics during the Cenozoic Era (65.5 million years ago to the present)—in particular, during the past 25 million years. Grasses, the dominant plants of savannas, appeared only about 50 million years ago, although it is possible that some savanna-like vegetation lacking grasses occurred earlier. The South American fossil record provides evidence of a well-developed vegetation, rich in grass and thought to be equivalent to modern savanna, being established by the early Miocene Epoch, about 20 million years ago.

Climates across the world became steadily cooler during this period. Lower ocean surface temperatures caused a reduction in water evaporation, which led to a slowing of the whole hydrologic cycle, with reduced cloud formation and less precipitation. The vegetation of mid-latitude regions, lying between the wet equatorial areas and the moist, cool, temperate zones, was affected substantially.

The main regions in which savannas emerged in response to this long-term climatic change—tropical America, Africa, South Asia, and Australia—were already separated from each other by ocean barriers by this time. Plant migration across these barriers was inhibited, and the details of the emergence of savannas on each continent varied. In each region different plant and animal species evolved to occupy the new, seasonally dry habitats.

Savannas became much more widespread, at the expense of forests, during the long, cool, dry intervals—contemporaneous with the Pleistocene Ice Ages, or glacial intervals, of temperate regions—during the Quaternary Period (2.6 million years ago to the present). Studies of fossilized pollen in sediments from sites in South America, Africa, and Australia provide strong support for this view.

When humans first appeared, in Africa, they initially occupied the savanna. Later, as they became more adept at modifying the environment to suit their needs, they spread to Asia, Australia, and the Americas. Here their impact on the nature and development of savanna vegetation was superimposed on the natural pattern, adding to the variation seen among savanna types. The savannas of

The sun sets on a savanna in Kenya. © Digital Vision/Getty Images

the world currently are undergoing another phase of change as modern expansion of the human population impinges on the vegetation and fauna.

Savanna Climates and Soils

In general, savannas grow in tropical regions 8° to 20° from the Equator. Conditions are warm to hot in all seasons, but significant rainfall occurs for only a few months each year—around October to March in the Southern Hemisphere and April to September in the Northern Hemisphere. Mean annual precipitation is generally 800 to 1,500 mm (31 to 59 inches), although in some central continental locations it may be as low as 500 mm (20 inches). The dry season is typically longer than the wet season, but it varies considerably, from two to eleven months. Mean monthly temperatures are around 10 to 20 °C (50 to 68 °F) in the dry season and 20 to 30 °C (68 to 86 °F) in the wet season.

Savannas may be subdivided into three categories—wet, dry, and thornbush—depending on the length of the dry season. In wet savannas the dry season typically lasts three to five months, in dry savannas five to seven months, and in thornbush savannas it is even longer. An alternative subdivision recognizes savanna woodland, with trees and shrubs forming a light canopy; tree savanna, with scattered trees and shrubs; shrub savanna, with scattered shrubs; and grass savanna, from which trees and shrubs are generally absent. Other classifications have also been suggested.

In spite of their differences, all savannas share a number of distinguishing structural and functional characteristics. Generally they are defined as tropical or subtropical vegetation types that have a continuous grass cover occasionally interrupted by trees and shrubs and that are found in areas where bushfires occur and where main growth patterns are closely associated with alternating wet and dry seasons. Savannas can be considered geographic and environmental transition zones between the rainforests of equatorial regions and the deserts of the higher northern and southern latitudes.

The distinction between savannas and other major vegetation types such as tropical deciduous forests, scrublands, or grasslands is somewhat arbitrary. The variation from one to another occurs along a continuum, often without distinct boundaries, and the vegetation is dynamic and changeable. The tree component of savannas generally becomes more important as rainfall increases, but other factors such as topography, soil, and grazing intensity all exert influences in complex and variable ways. Dry season fires, fueled by dried grass, may kill some trees, especially the more vulnerable young saplings; therefore, their severity also greatly affects the nature of savanna vegetation. Because grazing and fire are strongly affected by human activities and have been for thousands of years, humans continue to have a controlling influence on the nature, dynamics, development, structure, and distribution of savannas in many parts of their global range.

Soil fertility is generally rather low in savannas but may show marked small-scale variations. It has been demonstrated in Belize and elsewhere that trees can play a significant role in drawing mineral nutrients up from deeper layers of the soil. Dead leaves and other tree litter drop to the soil surface near the tree, where they decompose and release nutrients. Soil fertility in the vicinity of trees is thereby enhanced in comparison with areas between trees.

An unusually large proportion of dead organic matter—approximately 30 percent—is decomposed through the feeding activities of termites. Thus, a significant proportion of released mineral nutrients may be stored for long periods in termite mounds where they are not readily available to plant roots. In savannas in Thailand it has been shown that soil fertility can be markedly improved by mechanically breaking up termite mounds and spreading the material across the soil surface. In Kenya old termite mounds, which are raised above the general soil surface, also provide flood-proof sites where trees and shrubs can grow, with grassland between them, forming the so-called termite savanna.

Soil factors are particularly important in large areas of relatively moist savanna in South America and Africa. Where soils are poor, and especially in areas prone to waterlogging in the rainy season due to flatness of the ground or a hardpan close to the surface that roots cannot penetrate, tree growth is not vigorous enough for a closed forest to develop. This is true even where the climate appears to be suitable for it; a more open savanna vegetation is the result.

The Biota of Savannas

The biota of savannas reflect their derivation from regional biotas; therefore, species vary between regions. The savannas of Asia and tropical America, unlike those of Africa and Australia, are best considered as attenuated rainforests, their natural biotas having strong affinities with those of the wetter environments nearer the Equator in the same regions. Trees in these savannas are usually deciduous, their leaves falling during the dry season. The African savanna biota is fundamentally a grassland assemblage of plants and animals, with the addition of scattered trees.

Vegetation profile of a savanna. Encyclopædia Britannica, Inc.

Flora

Different groups of plants are prominent in the savannas of different regions. Across large parts of the tropical American savannas, the most common broad-leaved trees belong to the genera *Curatella*, *Byrsonima*, and *Bowdichia*, their place being taken in some seasonally waterlogged sites by the palms *Copernica* and *Mauritia*. Grasses include species of *Leersia* and *Paspalum*. In Argentina the most common woody plant is the bean relative *Prosopis*.

In the drier regions of East Africa, species of *Acacia* and *Combretum* are the most common savanna trees, with thick-trunked baobabs (*Adansonia digitata*), sturdy palms (*Borassus*), or succulent species of *Euphorbia* being conspicuous in some areas. In the drier savannas in particular there is often a wide diversity of spiny shrubs. Among the most prevalent grasses are species of *Andropogon*, *Hyparrhenia*, and *Themeda*. In wetter savannas, *Brachystegia* trees grow above a 3-metre- (10 feet) tall understory of elephant grass (*Pennisetum purpureum*). The most common West African savanna trees are in the genera *Anogeissus*, *Combretum*, and *Strychnos*.

In India the savanna vegetation of most areas has been extensively altered by human activities, which also have expanded its range. Where they have been least altered, Indian savannas commonly consist of thorny trees of *Acacia*, *Mimosa*, and *Zizyphus* growing over a grass cover consisting mainly of *Sehima* and *Dichanthium*.

At temperate latitudes in Australia the flora of the savanna resembles that of other types of sclerophyllous vegetation (thickened woody plants that have tough leaves with a low moisture content), neither fauna nor flora being of a distinctively savanna type. Most Australian savanna trees are evergreen, surviving the dry season not by dropping their leaves but by reducing water loss from them. The dominant trees of savannas in Australia and southern New Guinea are various species of *Eucalyptus*, with *Acacia*, *Bauhinia*, *Pandanus*, and other tall shrubs also being common. Baobabs (*Adansonia gregorii*) are the most common and conspicuous savanna trees in parts of northwest Australia. Tall spear grass (*Heteropogon*) or the shorter kangaroo grass (*Themeda*) dominates the understory of large areas of moist savanna. The prickly spinifex grasses (*Plectrachne*, *Triodia*) are prominent in more arid regions. Most trees and shrubs of the Australian savanna are markedly sclerophyllous. Small patches of monsoon rainforest and other types of vegetation occur locally within mainly savanna regions, surviving in places that have some degree of protection from the dry season fires.

Fauna

Savannas provide habitats for a wide array of animals, some of which foster the vegetation through grazing, browsing, pollinating, nutrient cycling, or seed dispersal. Many areas of savanna are managed today to maintain large grazing mammals, such as the native fauna of Africa or the cattle used for commercial production in large areas of Australia and South and Central America. Less spectacular but nevertheless very important are the small invertebrate animals; for example, grasshoppers and caterpillars are among the chief consumers of the understory foliage, and termites are significant consumers of dead plant matter, including wood.

Perhaps the best-known savanna fauna, because of its large mammals, is that of Africa. These large mammals basically are part of a grassland community, despite the presence of low trees in their environment. Most depend on the grass component of the vegetation either directly for their food, as do the herbivorous buffalo, zebra, gnu, hippopotamus, rhinoceros, and antelope, or indirectly, as is true of the carnivores or scavengers that feed primarily on these herbivores. Only a small number, including the giraffe and elephant, rely to a significant extent on foliage or fruit from the often thorny trees.

Large animals are uncommon in Australian savannas and are represented mainly by several species of the family Macropodidae, such as kangaroos and wallabies. However, in this region a wide variety of very large mammals and reptiles became extinct several thousand years ago, after the first arrival of humans. Their place today is taken by animals, both domesticated and feral, that have been introduced by humans: mainly cattle but also horses and, more locally, camels, donkeys, and the Asian water buffalo (*Bubalus bubalis*).

The Population and Community Development and Structure of Savannas

Savanna plants annually experience a long period in which moisture is inadequate for continued growth. Although the aboveground parts of the shallow-rooted grasses quickly dry out and die, the more deeply rooted trees can tap moisture lying farther beneath the surface longer into the dry season. Grasses grow rapidly when moisture is available but die back when it is not, surviving long, dry periods as dormant buds close to the soil surface. Sandy soils, which supply abundant moisture during rainy periods but which dry out almost completely in the absence of rain, favour the grassy component of savannas. Trees, on the other hand, require water in at least small amounts at all seasons even if they drop their leaves; deep soil layers supply this need. Trees in savannas are favoured by stony soils, which allow deep penetration by roots but which are less favourable to grasses. Nevertheless, especially toward the end of the dry season, many trees may lose their leaves to reduce transpirational loss of water, even though the leafless branches of some species carry open flowers. Soil, therefore, exerts some control over the nature of savanna vegetation, particularly in the drier parts of its distribution where sandy soils support grass-rich savanna with few trees and coarser, deeper soils support more tree-rich savanna with a smaller grass component.

Through the grazing pressure they exert, animals also can alter the balance between woody plants and grasses in a savanna—in either direction, depending on their feeding habits. Grass-eating mammals may overgraze and push the grass component of the vegetation toward local extinction; however, even high populations of these creatures cannot eliminate woody plant species, whose upper branches are out of their reach. Subsequent regeneration will favour the woody plants, which will become denser and shift the profile of the vegetation from savanna to forest. Other herbivores can have the reverse effect if their populations increase. For example, a steady rise in the elephant population between 1934 and 1959 in Virunga National Park, Congo (Kinshasa), led to an increase in the destruction of woody plants and transformed a heavily wooded savanna into a grass savanna with very few trees. An imbalance in favour of the tree components of savanna vegetation may also reduce the number and intensity of fires that would have destroyed many woody plants. Such bush encroachment commonly renders grazing land virtually useless; it is a widespread problem in drier parts of savanna lands in such places as Venezuela, India, and Australia.

Animals of savannas have adapted to surviving the seasonal variations in their food supply. Many birds and—especially in Africa—many mammals are seasonal migrants, occupying savannas during and immediately after the wet season when vegetation is lush and food abundant; they move elsewhere as the green parts of the plants disappear later in the dry season. The seasonal contrast in availability of plant food is less marked belowground where roots, tubers, and other subterranean organs commonly make up a large proportion of the total plant biomass—for example, up to four times as much as the aboveground component has been found below ground in some West African study sites, especially in the dry season. It is not surprising, therefore, that most invertebrate animals of the savanna—especially termites but also many other arthropods and earthworms—spend most of their lives underground.

Fire is an important ingredient in savanna ecosystems in all regions. Fires are started naturally by lightning strikes, but in most regions humans are now the greatest cause of savanna burning. Fire primarily consumes grasses, leaf litter, and other dead plant material that quickly dries out after the rains are over. Savanna trees commonly display a thick, corky bark that helps protect their trunks—at least once they have reached a certain size—from fire injury. While fires are important in the creation and maintenance of savanna vegetation in all regions, some disagreement exists concerning the extent to which fire should be considered a natural phenomenon, as well as to what extent it should be interpreted as the main factor responsible for the distribution and character of savanna vegetation.

Fires burn annually in savannas in all regions, nowhere more so than in Australia, the continent with the most fire-prone vegetation. In Australia humans have been lighting fires in savanna regions for at least 50,000 years. These fires have traditionally been lit for many reasons: to keep the country open and easily crossed; to reveal and kill small, edible animals such as lizards, turtles, and rodents; to create areas that later will develop a cover of fresh, green grass, which will attract wallabies and other game; and to encourage plants that produce edible tubers. Fires early in the dry season are less hot and destructive than fires that occur later in the season. They are sometimes employed to provide a firebreak around patches of fire-sensitive rainforest that inhabitants may want to protect for religious or utilitarian reasons. However, early fires may have ecological drawbacks, especially in areas intended for grazing. In these areas fires that burn late in the dry season are less detrimental to new grass growth.

The effect of fire on the vegetation is great. Some plants can survive fire. For example, some have buds located underground or beneath thick bark that provides fire protection; from these shielded structures regeneration quickly takes place. Other plants are able to reproduce effectively from seeds shed onto the fire-scorched ground in the wake of wildfire. Such plants benefit from burning and become more abundant than the fire-sensitive plants that occur in areas of frequent burning. Foremost among these plants are trees in the genus *Eucalyptus*, which contains many species that dominate most areas of Australian savanna. Some trees in Australian savanna areas, such as the cypress pine (*Callitris*), have been shown to be highly drought tolerant, albeit fire sensitive. Were it not for frequent fires, they would be able to grow over wide areas. Today *Callitris* is restricted to sites such as gorges and rocky outcrops where there is some protection from fire.

Similar patterns are recognizable in other regions. For example, in northern Nigeria thickets comprising a few fire-sensitive rainforest tree species from genera such as *Diospyros*, *Ficus*, and *Tamarindus* grow on rocky knolls lacking grass. These rocky "islands," protected from fire and cattle, are surrounded by expanses of grazed and frequently burned savanna. Where plots of African savanna vegetation are protected from being burned, they tend to revert quickly to deciduous forest.

Savannas are also affected by the overuse of woody plants for fuel. Together with grazing and cultivation, this leads to overall depletion of the vegetative cover, both the grassy and the woody components. Often a subsequent acceleration of soil erosion occurs. Such processes are associated, in densely settled savanna areas such as Africa north of the Equator, with the type of land degradation called desertification.

The Biological Productivity of Savannas

Savannas have relatively high levels of net primary productivity compared with the actual biomass (dry mass of organic matter) of the vegetation at any one time. Most of this productivity is concentrated into the period during and following the wet season, when water is freely available to the plants; at this time savanna productivity can rival or exceed that of forests. Values for the above ground biomass at its seasonal maximum range from 0.5 to 11.5 metric tons per hectare in drier regions (the higher values being recorded in years of sufficient rainfall) to 5.5 to 20.8 metric tons per hectare in more humid regions. Belowground biomass values have been measured less often but are typically as large as or larger than the aboveground values. Primary productivity is less easily evaluated, but rates of 3.6 metric tons of dry matter per hectare per year have been recorded in Senegal, a dry part of West Africa, and values of 21.5 to 35.8 metric tons per hectare per year in humid areas farther south. In India a range of values has been obtained for different savannas, from as low as 1.6 metric tons per hectare per year in drier areas to as high as 45.5 metric tons in wetter areas.

Furthermore, the quality of the vegetation as food for animals is generally high. A large proportion—ranging from 15 percent to more than 90 percent—is grass, which is palatable and digestible, especially by comparison with the woody vegetation that dominates forest growth. Grass foliage also contains far fewer unpalatable compounds than do most tropical forest tree leaves and so is more readily eaten and digested. Many shrubs and trees in savannas have leaves that are eaten by browsing mammals as well as invertebrates. Seeds and underground organs provide important dry-season foods for many animals.

Dried grass and dead wood in savannas are quickly decomposed, primarily by termites, or burned, releasing mineral nutrients to be reused in subsequent production.

CHAPARRAL

Chaparral is composed of broad-leaved evergreen shrubs, bushes, and small trees usually less than 2.5 metres (about 8 feet) tall; together they often form dense thickets. Chaparral is found in regions with a climate similar to that of the Mediterranean area, characterized by hot, dry summers and mild, wet winters. The name chaparral is applied primarily to the coastal and inland mountain vegetation of southwestern North America; sometimes it takes the place of a more general term, Mediterranean vegetation, which denotes areas of similar vegetation around the Mediterranean Sea, at the southern tip of Africa, in southwestern Australia, and in central South America.

Sages and evergreen oaks are the dominant plants in North American chaparral areas that have an average yearly rainfall of about 500 to 750 mm (20 to 30 inches). Areas with less rainfall or poorer soil have fewer, more drought-resistant shrubs such as chamise and manzanita. Chaparral vegetation becomes extremely dry by late summer. The fires that commonly occur during this period are necessary for the germination of many shrub seeds and also serve to clear away dense ground cover, thus maintaining the shrubby growth form of the vegetation by preventing the spread of trees. Chaparral returns to its prefire density within about 10 years but may become grassland by too frequent burning.

Deer and birds usually inhabit chaparral only during the wet season (the growth period for most chaparral plants), and move northward or to a higher altitude as food becomes scarce during the dry season. Small, dull-coloured animals such as lizards, rabbits, chipmunks, and quail are year-round residents. New chaparral growth provides good grazing for domestic livestock, and chaparral vegetation also is valuable for watershed protection in areas with steep, easily eroded slopes.

This rapid nutrient turnover helps explain the relatively high productivity and therefore the diverse and abundant faunas typical of savannas.

TUNDRAS

Tundras are areas of treeless, level, or rolling ground that occur in polar regions (Arctic tundra) or on high mountains (alpine tundra), characterized by bare ground and rock or by such vegetation as mosses, lichens, small herbs, and low shrubs.

The plant life of tundras tends to be greenish brown in colour, and species succession takes place slowly. The foggy tundras found along coastal areas produce matted and grassy swards. Algae and fungi are found along rocky cliffs, and rosette plants grow in rock cornices and shallow gravel beds. In the drier inland tundras, spongy turf and lichen heaths develop.

Tundra climates vary, the most severe being in the Arctic regions where temperatures fluctuate from 4 °C (40 °F) at midsummer to -32 °C (-25 °F) during the winter months. Alpine tundra has a more moderate climate, with cool summers and moderate winters (rarely falling below -18 °C [0 °F] in winter). The freezing climate of the Arctic produces a layer of permanently frozen soil, called the permafrost, which can reach soil depths of between 90 and 456 metres (300 and 1,500 feet). An overlying layer of soil alternates between freezing and thawing with seasonal variations in temperature. The permafrost layer exists only in Arctic tundra, but both Arctic and alpine tundras have a freeze-thaw layer.

Because Arctic tundras receive extremely long periods of daylight and darkness (lasting between one and four months), biological rhythms tend to be adjusted more to variations in temperature than to the amount of sunlight available for photosynthesis.

The Arctic Tundra

Arctic tundra covers about one-tenth of the total surface of the Earth. Its southern boundary meets the northernmost timberline, where boggy soils are threaded with numerous streams and lakes. Precipitation is less than

Forests and Grasslands

Cottongrass growing on the Arctic tundra in Alaska. © Photos.com/Jupiterimages

38 cm (15 inches) annually, and the sparse vegetation has a growing season between two and four months long. Most of the biological activity is confined to the freeze-thaw layer, because the softer soils of spring through autumn (thaw periods) allow animals to burrow, plant roots to extend down, and organic matter to decompose into food for microorganisms. Coastal tundras are dominated by mosses, sedges, and cotton grass. On more elevated sites, like hummocks, the soil is peatier and supports low willows, grasses, and rushes. Sunflower plants and legumes of various kinds thrive along the sandy banks of streams and lakes. Many of the plant species are perennials that flower within a few days of maturity, after the snows have begun to melt. They may germinate as soon as four to six weeks after maturing.

Animals common in Arctic tundras are the polar bear, Arctic fox, Arctic wolf, Arctic hare, and Arctic weasel. Many of these develop a white coat during the winter months as camouflage against the snow and ice. Large herbivores such as caribou, musk-oxen, and reindeer are adapted for the cold by virtue of their bulky bodies, since the lowered ratio of body surface area to mass (that is, to heat-producing tissue) reduces heat loss to the outside. Lemmings are an important species in the Arctic tundra. They remain active throughout the long winters, burrowing under the snow to feed on the roots of grasses and sedges. The accumulation of manure around their burrows adds nitrogen and other nutrients to the soil, stimulating plant growth.

Insects like mosquitoes and black flies, common to Arctic tundras, have adapted darkly coloured bodies to absorb as much heat from sunlight as possible. Many geese and other tundra birds are migratory, remaining in the tundra only long enough to nest and molt. Birds of prey (such as jaegers and snowy owls) and predatory animals (such as wolves and foxes) fluctuate in population levels according to the availability of their prey, particularly lemmings. In general, the food web in the tundra is simple and is easily subject to imbalance if a critical species fluctuates rapidly in population.

The Alpine Tundra

Alpine tundras begin above the timberline, either on gentle slopes where the soil has developed large meadow areas or on windswept slopes where cushion plants dominate. Annual precipitation is higher than in Arctic tundra; blinding snowstorms, or whiteouts, obscure the landscape during the winter months, and summer rains can be heavy. The stratification of the soil and the inclination of the alpine slopes allow for good drainage, however. Alpine tundra is dominated by shrubs and herbs; willows are common along streams or where snowdrifts are deep, as in basins or on the lee side of rock ridges. In the higher mountains, where the climate is more severe than along lower slopes, only lichens and mosses can survive.

Animal species are limited and only partially adapted to their wintry environment. Many enter into vertical migration patterns according to seasonal changes. Mountain sheep, ibex, chamois, wildcats, and many birds descend to warmer slopes to seek food in the winter. Some animals, like marmots and ground squirrels, consume large amounts of vegetation in the summer and early autumn and hibernate during the winter. Others, like rabbits, forage for what they can find in the snow; pikas and voles store large amounts of hay for winter feeding.

NOTABLE GRASSLANDS OF THE WORLD

Large grassland regions occur on most of the world's continents. The veld in southern Africa and the Llanos of South America are examples of such vast regions. Smaller grassland zones, such as Africa's Serengeti and North America's Buffalo Gap and Oglala National Grasslands, serve as valuable protected areas in the countries in which they appear.

Buffalo Gap National Grassland

Buffalo Gap National Grassland is a prairie grassland region of southwestern South Dakota, U.S. It covers an area of some 2,400 square km (925 square miles) of scattered land parcels and is divided into two districts. The eastern district, headquartered in Wall, runs along the northern border of the Pine Ridge Reservation of the Oglala Sioux and almost completely surrounds the northern part of Badlands National Park. The western district, headquartered in Hot Springs, extends south along the western border of the Pine Ridge Reservation and occupies the southwestern corner of the state. Oglala National Grassland in Nebraska adjoins it to the south. Established in 1960, it is administered as part of Nebraska National Forest.

Buffalo Gap's climate is semiarid, with fluctuating periods of precipitation and drought. The land is mostly flat, thus providing a wide-open and seemingly endless vista. The openness allows the almost constant wind to blow virtually unimpeded. Habitat is primarily midgrass prairie with areas of tallgrass prairie, shortgrass prairie, badlands, wetlands (mainly artificial), and woody patches along streams. Wildlife includes pronghorn, kangaroo rats, prairie dogs, coyotes, badgers, jackrabbits, deer, bison, bighorn sheep, and a wide variety of birds. The highly endangered black-footed ferret has been reintroduced. Nearly 50 species of grasses and hundreds of species of wildflowers grow in the grassland. Fossils are often found in the badlands, as well as in the grasslands; notable is a large cache of bison bones found just northwest of Crawford. Hunting and fishing are popular activities.

The visitors' centre, located in Wall, has historical displays, as well as specimens of rocks, minerals, and fossils found in the national grassland areas. The Black Hills region—which includes Black Hills National Forest, Custer State Park, Wind Cave National Park, Mount Rushmore National Memorial, and Jewel Cave National Monument—is north and west of Buffalo Gap. Cattle grazing on the grassland contributes to the economies of local communities.

The Llanos

The Llanos, which is the Spanish word for "plains," are wide grasslands stretching across northern South America and occupying western Venezuela and northeastern Colombia. The Llanos have an area of approximately 570,000 square km (220,000 square miles), delimited by the Andes Mountains to the north and west, the Guaviare River and the Amazon River basin to the south, and the lower Orinoco River and the Guiana Highlands to the east.

The elevations of the Llanos, rising from the Llanos Bajos ("Low Plains") west of the Orinoco River to the Llanos Altos ("High Plains") below the Andes, rarely exceed 300 metres (1,000 feet). The Llanos Altos form extensive platforms between rivers and rise 30 to 60 metres (100 to 200 feet) above the valley floors. The Llanos are drained by

Horses being watered on the Llanos, in eastern Colombia. © Victor Englebert

the Orinoco and its western tributaries, including the Guaviare, Meta, and Apure rivers. Annual precipitation is concentrated between April and November and ranges from 1,100 mm (45 inches) in Ciudad de Nutrias in the central plains to 4,570 mm (180 inches) in Villavicencio near the Andes. Mean daily temperatures in the Llanos exceed 24 °C (75 °F) throughout the year.

Most of the Llanos is treeless savanna that is covered with swamp grasses and sedges in the low-lying areas and with long-stemmed and carpet grasses in the drier areas. Much of the Llanos Bajos is subject to seasonal flooding. Trees are concentrated along rivers and in the Andean piedmont; trees scattered on the open savanna include scrub oak and dwarf palm. Most mammals nest in the gallery forests and feed on the grassland; among these are included several species of deer and rabbit, as well as the anteater, armadillo, tapir, jaguar, and capybara, which is the world's largest living rodent.

The raising of cattle has long been the mainstay of the Llanos' economy, since Spanish colonial days. Since the 1950s there has also been considerable small farming. The economic importance of the region has been greatly enhanced by the oil fields in the Venezuelan Llanos at El Tigre and Barinas.

The Nyika Plateau

The Nyika Plateau is a high grassy tableland in northern Malawi. It is a tilted block extending from the Mzimba Plain northeast to the edge of the Great Rift Valley and Lake Nyasa. Its undulating surface is covered with montane grassland and patches of evergreen forest (including the southernmost occurrence of the Mulanje tree [*Juniperus procera*]) and is marked by occasional peaks and ridges (Nganda, 2,606 metres [8,551 feet]; Vitumbi, 2,527 metres [8,291 feet]). A beveled surface at 1,676 metres (5,500 feet) rings the plateau. High rainfall generates the formation of perennial rivers in *madambo* (broad grass-covered depressions) and their flow through deep erosional valleys. The main streams include the North Rumphi, Chelinda (Rumphi), Runyina, and North Rukuru rivers. The higher soils are too poor for cultivation and the plateau is virtually uninhabited. Nyika National Park (founded 1965) covers 3,134 square km (1,210 square miles) of the plateau surface and supports antelope, zebra, lion, and trout.

Oglala National Grassland

The Oglala National Grassland is a region of federally recognized prairie grasslands of northwestern Nebraska, U.S. The designated national grassland covers a noncontiguous area of some 390 square km (150 square miles) in the

Nebraska panhandle, including scattered parcels of land in Sioux and Dawes counties bordering the states of South Dakota and Wyoming. Headquarters are in Chadron. Established in 1960, it is administered as part of the Nebraska National Forest. Buffalo Gap National Grassland in South Dakota adjoins it to the north.

The landscape of Oglala National Grassland includes badland areas where toadstool formations (rocks eroded into mushroomlike shapes) and fossil deposits are found. The Hudson–Meng Bison Bonebed, discovered in the 1950s, contains the remains of hundreds of prehistoric bison that all died at the same time from an unknown cause some 10,000 years ago. Stone artifacts of the Paleo-Indian Alberta culture have been found in association with the bones, leading to a theory that the bison were killed by nomadic hunters; subsequent scholarship, however, deemed it more likely that the bison died of a natural cause such as suffocation during a prairie fire.

Oglala National Grassland contains ideal rangeland for cattle, and grazing is an important use of the land. Pronghorn are common, and the grassland is one of the state's most popular areas for hunting. Other animals include deer, wild turkeys, grouse, foxes, burrowing owls, and prairie dogs. Ponds provide habitat for waterfowl and opportunities for fishing. The National Grasslands Visitor Center in Wall, S.D., features exhibits on the history and the flora and fauna of the grassland.

Serengeti National Park

The Serengeti National Park is a national park and wildlife refuge that occurs on the Serengeti Plain in north-central Tanzania. It is partly adjacent to the Kenya border and is northwest of the adjoining Ngorongoro Conservation Area. It is best known for its huge herds of plains animals (especially gnu [wildebeests], gazelles, and zebras), and it is the only place in Africa where vast land-animal migrations still take place. The park, an international tourist attraction, was added to the UNESCO World Heritage List in 1981.

The park was established in 1951 and covers 14,763 square km (5,700 square miles) of some of the best grassland range in Africa, as well as extensive acacia woodland savanna. With elevations ranging from 920 to 1,850 metres (3,020 to 6,070 feet), the park extends 160 km (100 miles) southeast from points near the shores of Lake Victoria and, in its eastern portion, 160 km (100 miles) south from the Kenya-Tanzania border. It is along the "western corridor" to Lake Victoria that many of the park's animals migrate. Within the area are nearly 1,300,000 gnu, 60,000 zebras, 150,000 gazelles, and numerous other animals. During the wet season, from November to May, the herds graze in the southeastern plains within the park. In late May or June one major group moves west into the park's woodland savanna and then north into the grasslands just beyond the Kenya-Tanzania border, an area known as the Mara (Masai Mara National Reserve). Another group migrates directly northward. The herds return to the park's southeastern plains in November, at the end of the dry season.

In addition to more than 35 species of plains animals, there are some 3,000 lions and great numbers of spotted hyenas, leopards, rhinoceroses, hippopotamuses, giraffes, cheetahs, and baboons. Crocodiles inhabit the marshes near the Mara River. More than 350 species of birds, including ostriches, vultures, and flamingos, have also been recorded.

Elephants, which were not found in the Serengeti until 30 years ago, moved into the park as human populations and agricultural developments increased outside its borders; the local elephant population is estimated at some 1,360. The last of the Serengeti's wild dogs disappeared in 1991, but there are some 30,000 domestic dogs in the area; it is possible that unvaccinated domestic dogs spread rabies to the wild dogs, resulting in their local extinction. An epidemic of canine distemper caused the deaths of nearly one-third of the area's lions in 1994. The killing of elephants for their ivory tusks, the slaughter of the now virtually extinct black rhinoceros for its horn (which is prized in Yemen for dagger handles), and the poaching of game animals for meat—an estimated 200,000 a year—are major threats.

The first systematic wildlife population survey in the area was undertaken by the German zoologist Bernhard Grzimek in the late 1950s. The park's headquarters are near its centre, at Seronera, where the Seronera Wildlife Research Centre (established as the Serengeti Research Institute, 1962) is also based.

Veld

Veld, which is the Afrikaans word for "field," is the name given to various types of open country in Southern Africa that is used for pasturage and farmland. To most South African farmers today the "veld" refers to the land they work, much of which has long since ceased to be "natural."

Various types of veld may be discerned, depending upon local characteristics such as elevation, cultivation, and climate. Thus, there is a high veld, a middle veld, a low veld, a bush veld, a thorn veld, and a grass veld. The boundary between these different varieties of veld is frequently vague, and all of them are usually referred to with the general term *veld* by the local inhabitants. For convenience, its major regions—Highveld, Middleveld, and Lowveld—are distinguished on the basis of elevation.

Physiography

The Highveld comprises most of the high-plateau country of Southern Africa. Except in Lesotho, where it extends well above 2,500 metres (8,200 feet) and even above 3,350 metres (about 11,000 feet) in places, all of it lies between 1,200 and 1,800 metres (about 4,000 and 6,000 feet) above sea level. The South African part of the region is bounded to the east and south by the Great Escarpment, which consists of the Drakensberg and Cape ranges, and by the Lesotho Highlands. Its less clearly defined northern and western boundaries coincide roughly with the 4,000-foot contour. Most of it is underlain by sedimentary strata of the Karoo System (or Karoo Super Group), dating from about 345 to 190 million years ago, and to older pre-Karoo material. Among these are coal-bearing strata. These materials have been eroded over a long period of time to produce generally flat plains, dissected occasionally by deeply carved valleys and including relict mountains and scattered steep-sided hills called kopjes, or koppies. The Highveld plains are thought to have been created by pedimentation, in which the areas around resistant rock are eroded away, leaving mountains of low relief and kopjes. Large areas of the western part of the region are also covered by "pans," which are shallow and ephemeral lakes, often with salty crusts; these are found especially in several provinces in South Africa.

In Zimbabwe the Highveld coincides roughly with the region lying on either side of the central watershed. Like the Highveld of South Africa, it has a remarkably even surface, broken only by kopjes and low ridges. Throughout the Highveld, soils tend to be thin, poor, and powdery and thus easily carried away by both wind and water erosion.

The Middleveld is the name given in South Africa to a vast and geologically complex region that lies in the region north of Pretoria, in the Northern Cape province, and in Namibia. Its boundaries are not as well defined as are those for the Highveld, but generally it lies at an altitude between 600 and 1,200 metres (2,000 and 4,000 feet) above sea level. In Zimbabwe to the northeast, the Middleveld also consists of the land lying roughly between 600 and 1,200 metres. Most of the Middleveld is underlain by Precambrian rocks that have been exposed by erosion. In northern South Africa it is underlain by the unique Bushveld complex, with its wealth of rare minerals. As is the case in the Highveld, the uniformity of the relief is broken by relict mountains and by kopjes. Pans are numerous, especially in the western areas. Middleveld soils are generally thin and poor.

The Lowveld is the name given to two areas that lie at an elevation of between 150 and 600 metres (500 and 2,000 feet) above sea level. One area is in the South African provinces of Mpumalanga and KwaZulu-Natal and parts of Swaziland, and the other is in southeastern Zimbabwe. Both are underlain largely by the soft sediments and basaltic lavas of the Karoo System and by loose gravels. They have been extensively intruded by granites. Other resistant metamorphic rocks also occur; these commonly appear as low ridges or what seem to be archipelagoes of island mountains. The higher western margins of both areas testify to the degree of erosion resulting from the flow of rivers running east or southeast.

The soils of the Lowveld are more varied than those of the other veld regions. Along their higher and wetter western sides, they tend to be deep, leached (percolated by water), acidic, porous, and well drained. In the lower-lying and drier central and eastern portions, they tend to be shallow, but they are more fertile and retain moisture better.

Climate

The climate of the veld is highly variable, but its general pattern is mild winters from May to September and hot or very hot summers from November to March, with moderate or considerable variations in daily temperatures and abundant sunshine. Precipitation mostly occurs in the summer months in the form of high-energy thunderstorms.

Over most of the South African Highveld, the average annual rainfall is between 380 and 760 mm (15 and 30 inches), decreasing to about 250 mm (10 inches) near the western border and increasing to nearly 1,000 mm (40 inches) in some parts of the Lesotho Highlands; the South African Lowveld generally receives more precipitation than the Highveld. Temperature is closely related to elevation. In general, the mean July (winter) temperatures range between 7 °C (45 °F) in the Lesotho Highlands and 16 °C (60 °F) in the Lowveld. January (summer) temperatures range between 18 °C (65 °F) and 27 °C (80 °F).

In Zimbabwe the precipitation averages around 760 to 900 mm (30 to 35 inches) on the Highveld, dropping to less than 380 mm (15 inches) in the lowest areas of the Lowveld. Temperatures are slightly higher than in South Africa.

Over the entire veld, seasonal and annual average rainfall variations of up to 40 percent are common. Damaging drought afflicts at least half the area about once every three or four years. Everywhere the average number of hours of annual sunshine varies from 60 to 80 percent of the total amount possible.

Plant Life

The veld regions support an enormous variety of natural vegetation. No particular species is ubiquitous, and many are highly localized. Grassveld is the characteristic vegetation of the South African Highveld, dominated by species of red grass. Where the red grass grows on well-drained, fertile soils subject to comparatively light rainfall, it tends to be sweeter (and is consequently called sweetveld) than elsewhere, where it is commonly called sourveld. Sweetvelds are more palatable to livestock than sourvelds, the latter being usable as fodder only in winter.

The drier South African Middleveld favours both red grass and drought-resistant species of grasses. These

Highveld grassland near Heidelberg, S.Af., southeast of Johannesburg. Gerald Cubitt—Bruce Coleman Ltd.

grasses are less luxuriant and the ground cover less complete than those of the Highveld. As the aridity increases to the west and north, the cover becomes sparser, and grassveld gradually loses ground to thornveld (consisting of such types as thorny acacias and aloes), dwarf, drought-resistant bushes, and desert scrub.

In Zimbabwe the Highveld and Middleveld consist of open woodland savanna dominated by leguminous, fire-resistant trees of the *Brachystegia* genus. Tall perennial grasses and flowering herbs, which readily catch fire during the dry season, occupy most of the open ground.

The Lowveld everywhere supports a parklike plant cover. In the higher areas the characteristic trees are acacia and marula, the latter bearing an intoxicating plumlike fruit. The open ground is dominated by red grass. In the lower areas, such as the Sabi and Limpopo river valleys, tufted finger grasses, euphorbias, and other succulents replace red grass; the acacias increase in number; and the mopane tree, the baobab, and the tall fan palm occur.

Animal Life

Mass slaughter, trophy hunting, and the encroachments of farmers and pastoralists have thinned out every major species of mammal and reptile and several species of birds in the veld. The South African and Zimbabwean governments have, however, set aside vast tracts of veld as wildlife reserves. Wildlife conservation efforts in Southern Africa have further been aided by the creation of transfrontier parks, which link nature reserves and parks in neighbouring countries to create large, international conservation areas that protect biodiversity and allow a wider range of movement for migratory animal populations. One such park is the Great Limpopo Transfrontier Park, which links Kruger National Park in South Africa with Limpopo

National Park in Mozambique and Gonarezhou National Park in Zimbabwe. The lion, leopard, cheetah, giraffe, elephant, hippopotamus, oryx, kudu, eland, sable antelope, and roan antelope survive only in or near such protected areas. The smaller mammals, most of the reptiles, and almost all of the birds—except the ostrich, which has virtually been eliminated from the veld—are still found in the wild.

CONCLUSION

Taken together, forests and grasslands cover approximately 70 percent of Earth's land area. Thus, it is no surprise that they are tremendous repositories of life and home to the overwhelming majority of the planet's species. These biomes contain many familiar plants and animals; however, they are also home to multitudes of undescribed species.

Forested areas are typically adjacent to and interwoven with grassland areas. Combination ecosystems, such as forested grasslands and glades (that is, grassy clearings within forests), are common especially where grasslands and forests meet. In such areas, the structural differences between these biomes become apparent.

Forests are often characterized by tall vegetation with wide woody trunks and expansive leafy canopies. In most forests, shorter shrubby plants occur on the forest floor, soaking up sunlight that filters through the leaves of taller trees. Together, both tall and short plants, along with the soil they are rooted in, provide living space and food resources for large numbers of animals, fungi, and other organisms. Figuratively speaking, many forests have become worlds unto themselves; forest interiors are cooler, shadier, and more humid than the ecosystems that surround them. In addition, forest interiors are characterized by little, if any, wind.

Grasslands, on the other hand, tend to be drier than forests. Winds blow easily over vast areas of nonwoody plants called grasses. Compared to the trees and the shrubs of the forests, grasses tend to be shorter and thinner, and sunlight generally penetrates the vegetative cover down to the soil. This combination of wind and sunlight ensures that grasslands remain low-humidity environments. Although not as biologically diverse as forested areas, grasslands do contain specialized plants and animals capable of dealing with variable temperature and moisture conditions. In addition, many of the grasses have evolved to survive wildfires and take advantage of the scorched ground bereft of competing species.

Forests and grassland biomes are also home to most of the world's people. Building materials, game animals, crops, pharmaceuticals, tools, and many other useful items are derived from forest and grassland plants and animals. Since the dawn of civilization, people have converted vast areas of these biomes into croplands for agriculture, grazing lands for domestic animals, and built-up areas for living space. Such land use conversion continues today. Deforestation continues to plague tropical and temperate forests alike, and trees are removed from an area comparable to the size of Panama each year. As the human population grows and more and more people move to metropolitan areas, the spatial footprint of many cities and towns around the world also expands, forcing the conversion of adjacent woodlands and grasslands to croplands, roads, residential areas, and other human-dominated landscapes.

For other forms of life on Earth, habitat lost to deforestation and urban sprawl results in stressed populations. Although some species thrive in human-dominated landscapes, most do not. When faced with habitat loss, plants, animals, and other organisms whose populations are limited to small geographic ranges often suffer dramatic population declines or become extinct. In tropical areas some of these species have died out before they could be described by scientists.

To prevent the extinction of more species, the protection and effective management of these valuable biomes must become a priority. The concepts of multiple use and sustainable yield pioneered in forestry are valuable tools for reducing the risk of species extinction. Multiple use provides a mechanism with which competing interests can share a resource, whereas sustainable yield ensures the perpetuity of the resource. In addition, the application of other forestry practices, such as replanting harvested areas and lowering the intensity of human disturbances, could allow people to harvest the amenities of forests and grasslands without destroying them.

APPENDIX

AVERAGE NET PRIMARY PRODUCTION OF THE EARTH'S MAJOR HABITATS

HABITAT	NET PRIMARY PRODUCTION (GRAM PER SQUARE METRE PER YEAR)
Forests	
tropical	1,800
temperate	1,250
boreal	800
Other Terrestrial Habitats	
swamp and marsh	2,500
savanna	700
cultivated land	650
shrubland	600
desert scrub	70
temperate grassland	500
tundra and alpine	140
Aquatic Habitats	
algal beds and reefs	2,000
estuaries	1,800
lakes and streams	500
continental shelf	360
open ocean	125

Source: Adapted from Robert E. Ricklefs, Ecology, 3rd edition (1990), by W.H. Freeman and Company, used with permission.

Glossary

abscission layer A fragile plant tissue layer at the bottom of a leaf's stalk that allows the leaf to separate easily from the plant.
arboreal Relating to or living in trees.
bedrock The solid rock beneath layers of soil.
biomass The total quantity of organic matter, including all the plants and animal species, in an area at a certain moment.
chlorophyll A green pigment found in plants that enables the process of photosynthesis to take place.
conifer Trees and shrubs, most of which are evergreens and have needle-shaped or scaly leaves.
durian A flowering plant that has a spherical fruit with an eight-inch diameter, a custard-like flavor, and an overpowering smell.
ecosystem An ecological unit made up of organisms and the environment they live in.
epiphytes Plants that grow on the surface of other plants but get nutrients and water from the air.
estuaries A partly enclosed body of water near a coast where saltwater and fresh water mix.
evapotranspiration When both evaporation and transpiration from plants cause water to be lost from soil.
humus A kind of rich organic soil made up of decomposed plant and animal matter.
hydrologic cycle The circulation of water through the Earth's atmosphere, from evaporation, condensation, precipitation to runoff.
insoluble A substance that cannot be dissolved in a liquid.
inundated Flooded.

lianas Vines rooted in the earth that climb up and wrap around other plants. Lianas are common in tropical forests.

microclimate A climatic condition that takes place in a small area.

microorganisms A diverse group of minute, simple life-forms that includes bacteria, archaea, algae, fungi, protozoa, and viruses.

monsoon A large-scale wind system that seasonally reverses its direction.

niche The interactions, such as competition, mutualism, parasitism, and predation, of a species with the other members of its community.

refugia A region where, in a time of large-scale climate change such as an ice age, the climate remains relatively stable, thus providing a haven for plants and animals that can repopulate the land when the climate becomes more hospitable.

saprophytic Referring to a type of plant life that grows on dead or decayed material.

stomates Tiny openings in leaves and stems that allow for the exchange of gases between the open air and the inside of the leaf.

symbiotic A close, but not necessarily mutually beneficial, relationship between two organisms of different species.

transpiration The release of water vapor from a living organism, such as a tree, through a membrane (plants) or pores (animals).

BIBLIOGRAPHY

TROPICAL RAINFOREST

P.W. Richards, *The Tropical Rain Forest* (1952), is perhaps the best early account, containing a mine of useful information. Kathlyn Gay, *Rainforests of the World: A Reference Handbook* (1993), describes the interaction of rainforests and climate. Regional accounts of value are T.C. Whitmore, *Tropical Rain Forests of the Far East*, 2nd ed. (1984); and Paul Adam, *Australian Rainforests* (1992).

TEMPERATE FOREST

Broad descriptions of temperate forests can be found in J.D. Ovington (ed.), *Temperate Broad-Leaved Evergreen Forests* (1983); and E. Röhrig and B. Ulrich (eds.), *Temperate Deciduous Forests* (1991). J.R. Packham et al., *Functional Ecology of Woodlands and Forests* (1992), provides a general account of forest ecology, with particular reference to European deciduous forests. M. Numata (ed.), *The Flora and Vegetation of Japan* (1974), describes the temperate broad-leaved evergreen and deciduous forests of East Asia.

BOREAL FOREST

Herman H. Shugart, Rik Leemans, and Gordon B. Bonan (eds.), *A Systems Analysis of the Global Boreal Forest* (1992), covers ecosystem processes, forest patterns in space and time, and computer models, including chapters on the Eurasian boreal forest, tree and shrub reproduction, and fire in the boreal forest. *The Plant Cover of Sweden* (1965), is the most complete reference on the original condition of the boreal forest of Sweden, focusing on the landscape pattern of native vegetation and on plant indicators of various forest regions and ecosystem types. Deborah L. Elliott-Fisk, "The Boreal Forest," in Michael G. Barbour and William Dwight Billings (eds.), *North American Terrestrial Vegetation* (1988), pp. 33–62, is a basic reference on the forest types and communities found across the North American boreal region and includes the historical development of the forest and the ecological characteristics of individual tree species. K. Van Cleve et al. (eds.), *Forest Ecosystems in the Alaskan Taiga: A Synthesis of Structure and Function* (1986), looks at recent investigations of one of the best-studied parts of the boreal forest, describing both upland and floodplain ecosystems; an update in a special issue of *Canadian Journal of Forest Research*, vol. 23, no. 5 (May 1993), contains the results of a multidisciplinary research project on ecological succession of a productive river floodplain in central Alaska, including a detailed discussion on soil. J.S. Rowe, *Forest Regions of Canada* (1972), well illustrated, is the best reference on the overall distribution and classification of boreal forest types in Canada. James A. Larsen, *The Boreal Ecosystem* (1980), focuses on properties of the middle boreal forest of central Canada. Edward A. Johnson, *Fire and Vegetation Dynamics: Studies From the North American Boreal Forest* (1992), is a well-illustrated treatment of fire weather, fire behaviour, and fire effects in the North American boreal forest. Lennart Hansson (ed.), *The Ecological Principles of Nature Conservation: Applications in Temperate and Boreal Environments* (1992), also well illustrated, discusses the effects of forest management on biodiversity, focusing particularly on the boreal forest of northern Europe.

FORESTRY

Forestry textbooks usually deal with a specialized aspect of the science, often in one geographic region. William M. Harlow, Ellwood S. Harrar, and Fred M. White, *Textbook of Dendrology, Covering Important Forest Trees of the United States and Canada*, 6th ed. (1979), provides descriptions of major tree species in the region. Forest ecology is the subject of K.A. Longman and J. Jeník, *Tropical Forest and Its Environment*, 2nd ed. (1987); J.J. Landsberg, *Physiological Ecology of Forest Production* (1986); and Herman H. Shugart, *A Theory of Forest Dynamics: The Ecological Implications of Forest Succession Models* (1984). For a general historical overview of forestry and related fields, see Richard C. Davis (ed.), *Encyclopedia of American Forest and Conservation History*, 2 vol. (1983).

Principles of forest management, worldwide in application, are systematically outlined by William A. Leuschner, *Introduction to Forest Resource Management* (1984); and Joseph Buongiorno and J. Keith Gilless, *Forest Management and Economics: A Primer in Quantitative Methods* (1987). Karl F. Wenger (ed.), *Forestry Handbook*, 2nd ed. (1984), is a reference book of data and methods in all aspects of forestry and allied fields. Grant W. Sharpe, Clare W. Hendee, and Wenonah F. Sharpe, *Introduction to Forestry*, 5th ed. (1986); and Charles H. Stoddard and Glenn M. Stoddard, *Essentials of Forestry Practice*, 4th ed. (1987), give a complete overview of modern multiple-use forestry. The requisite elements of forest inventory are detailed in Bertram Husch, Charles I. Miller, and Thomas W. Beers, *Forest Mensuration*, 3rd ed. (1982). Financial implications are studied in G. Robinson Gregory, *Resource Economics for Foresters* (1987). The leading international sources of statistics on forestry and timber output are the publications of the Food and Agriculture Organization of the United Nations: *World Forest Inventory* (irregular), *FAO Forestry and Forest Product Studies* (irregular); and the special reports *Forest Resources of Tropical Africa*, 2 vol. (1981), *Forest Resources of Tropical Asia* (1981), and *Forestry in China* (1982). Fundamental studies of the physical bases of forest distribution and yield are given in *World Resources 1986* (1986), a report prepared by the World Resources Institute and the International Institute for Environment and Development.

The theory and practice of raising and tending tree crops are treated in such manuals as Theodore W. Daniel, John A. Helms, and Frederick S. Baker, *Principles of Silviculture*, 2nd ed. (1979); and David M. Smith, *The Practice of Silviculture*, 8th ed. (1986), discussing temperate-zone forests. Details of handling seed and young stock are treated in R.L. Willan (comp.), *A Guide to Forest Seed Handling: With Special Reference to the Tropics* (1985); U.S. Department of Agriculture, *Woody-Plant Seed Manual* (1948); and Mary L. Duryea and Thomas D. Landis (eds.), *Forest Nursery Manual: Production of Bareroot Seedlings* (1984). Genetic improvement is discussed in Klaus Stern and Laurence Roche, *Genetics of Forest Ecosystems* (1974); and M.N. Christiansen and Charles F. Lewis (eds.), *Breeding Plants for Less Favorable Environments* (1982). The intensive culture of forest plantations is discussed in Bruce J. Zobel, Gerrit Van Wyk, and Per Stahl, *Growing Exotic Forests* (1987); and W.E. Hillis and A.G. Brown (eds.), *Eucalypts for Wood Production* (1978, reprinted 1984). Management of forest soils, a primary concern in intensive silviculture, is described in William L. Pritchett and Richard F. Fisher, *Properties and Management of Forest Soils*, 2nd ed. (1987); and Pedro A. Sanchez, *Properties and Management of Soils in the Tropics*

(1976). Manipulation of soil fertility and study of soil microorganisms are presented in G.D. Bowen and E.K.S. Nambiar (eds.), *Nutrition of Plantation Forests* (1984); Robert L. Tate, III, and Donald A. Klein (eds.), *Soil Reclamation Process: Microbiological Analyses and Applications* (1985); and J.C. Gordon and C.T. Wheeler (eds.), *Biological Nitrogen Fixation in Forest Ecosystems: Foundations and Applications* (1983). The care of tree crops in tropical jungles of both hemispheres is outlined in I.T. Haig, M.A. Huberman, and U. Aung Din, *Tropical Silviculture* (1958). A wide range of Asiatic conditions is discussed in Harry G. Champion and S.K. Seth, *General Silviculture for India* (1968).

The management of forests as watersheds and the impact of forestry activities on water quantity and quality are discussed in William E. Sopper and Howard W. Lull (eds.), *Forest Hydrology: Proceedings of a National Science Foundation Advanced Science Seminar* (1967); H.C. Pereira, *Land Use and Water Resources in Temperate and Tropical Climates* (1973); and K.W.G. Valentine, *Soil Resource Surveys for Forestry* (1986). Nutrient losses from disturbed watersheds and the potential for accelerated loss attributable to acid deposition are examined in F.E. Clark and T. Rosswall (eds.), *Terrestrial Nitrogen Cycles: Processes, Ecosystem Strategies, and Management Impacts* (1981); and S. Beilke and A.J. Elshout (eds.), *Acid Deposition* (1983).

Forest protection is treated in textbooks discussing specific hazards. Arthur A. Brown and Kenneth P. Davis, *Forest Fire: Control and Use*, 2nd ed. (1973), outlines fire dangers and methods of control. T.T. Kozlowski and C.E. Ahlgren (eds.), *Fire and Ecosystems* (1974); and Henry A. Wright and Arthur W. Bailey, *Fire Ecology, United States and Southern Canada* (1982), discuss the environmental interactions associated with fire. John S. Boyce, *Forest Pathology*, 3rd ed. (1961); and Robert O. Blanchard and

Terry A. Tattar, *Field and Laboratory Guide to Tree Pathology* (1981), give details of fungal diseases, climatic dangers, and airborne fume and salt damage. See also Carl F. Jordan (ed.), *Amazonian Rain Forests: Ecosystem Disturbance and Recovery* (1987). Insect pests are described in Alan A. Berryman, *Forest Insects: Principles and Practice of Population Management* (1986).

GRASSLAND

The grassland ecology of North America is discussed in Lauren Bowen, *Grasslands* (1985).

SAVANNA

A wide assortment of issues relating to the global distribution and ecology of savannas are addressed in François Bourlière (ed.), *Tropical Savannas* (1983). E.M. Lind and M.E.S. Morrison, *East African Vegetation* (1974), discusses the range of savannas and related vegetation types in East Africa. Patricia A. Werner (ed.), *Savanna Ecology and Management: Australian Perspectives and Intercontinental Comparisons* (1991), deals with the functioning, maintenance, and management of savannas, focusing upon the region in which they remain most intact and undamaged by human activities.

Index

A

agroforestry, 176–177
Amazon Rainforest, 106–112
angiosperms, 9–12, 19, 140–142
   animal life of, 110–112
      plant life of, 108–110

B

Bavarian Forest, 119–120
Belovezhskaya Forest, 120–121
Black Forest, 121–122
Boise National Forest, 122–123
boreal (taiga) forest, 1, 53, 75–105
   biological productivity of, 101–102
   biota of, 81–82, 86–97
   climate of, 79–83
   community structure of, 98–101
   deforestation of, 102–105
   distribution of, 77–79
   effects of human use and management on, 99–101
   fires in, 98–99
   origin of, 75–76
   soils of, 3, 83–86
bromeliads, 22–23
Buffalo Gap National Grassland, 211–212

C

chaparral, 206
cloud forests, 25
coniferous forests, 81–82, 143–144, 168
   structure of, 3–4

D

deciduous forests, temperate, 2, 3, 4, 53, 55–59, 60–61, 62–64, 69–73, 74, 144, 168
deciduous forests, tropical, 13, 14, 15
   biota of, 18–19
deforestation, 224
   of boreal forests, 102–105
   of deciduous forests, 55–56
   global impact of, 44–45
   of rainforests, 13–14, 44–51

E

ecotourism, 52
evergreen forests, 53–54, 58, 62, 64–67, 74, 82, 144–145, 168
extinctions of biota, 19, 225

F

fauna (animals)
   in the Amazon Rainforest, 110–112
   in boreal forests, 90–97
   of grasslands, 183–184, 185, 186
   in the Ituri Forest, 117–118
   of prairies, 192
   of savannas, 200
   in temperate forests, 67–68

in tropical rainforests, 23–25,
   33–40
of the tundra, 209–210
of veld, 222–223
flora (plants)
   in the Amazon Rainforest,
      108–110
   in boreal forests, 86–89
   of grasslands, 183, 184–186
   in the Ituri Forest, 116–117, 118
   of prairies, 191–192
   of savannas, 198–199
   in temperate forests, 61–67
   in tropical rainforests, 19–23,
      33–40
   of the tundra, 209, 210
   of veld, 221–222
forest fires, 166–173
forests
   classification and distribution
      of, 140–146
   notable forests of the world,
      106–132
   occurrence and distribution
      of, 142–146
   recreation in, 160–162
   wildlife in, 162–163
forestry
   agroforestry, 176–177
   in ancient world, 133–134
   classification and distribution
      of forests, 140–146
   development of U.S. policies,
      137–140
   and fire prevention and
      control, 166–173
   history of, 133–140
   insect and disease control,
      173–176

in medieval Europe, 134–135
   modern developments in,
      135–137
   and multiple-use concept, 147
   purposes and techniques of
      forest management, 146–177
   and range and forage, 159–160
   and recreation and wildlife,
      160–163
   and sustained yield, 147–149
   urban, 177
   watershed management and
      erosion control, 164–166
Franconian Forest, 123–124, 130

G

grasses, 189
grasslands
   biological productivity of,
      188–190
   biota of, 183–186
   climates and soils of, 182–183
   development and structure of,
      186–188
   notable grasslands of the
      world, 211–223
   origin of, 178–182
   temperate, 181, 182, 186–187
   tropical, 181, 182, 186–187
gymnosperms, 140–141

H

herbivory, 42–43

I

insect and disease control,
   173–176

Index

Ituri Forest, 106, 113–118
  climate and drainage of, 115–116
  physiography of, 113–115
  plant and animal life of, 116–118
  soils of, 116

L

Llanos, the, 212–214

M

mangrove forests, 5
monsoon forests, 5, 20
Mount Hood National Forest, 124–125

N

Nyika Plateau, 214

O

Oglala National Grassland, 214–215
Ozark–Saint Francis National Forest, 125–126

P

Panamanian rainforest
  aerial seed dispersal in, 27–28
  forest regeneration in, 34
  seed dispersal by animals in, 37–38
permafrost, 84–85, 99
Pinchot, Gifford, 136, 138, 147
prairies, 190–192

S

savannas, 192–206
  biological productivity of, 204–206
  biota of, 197–200
  climates and soils of, 195–197
  development and structure of, 201–204
  fires in, 202–204
  origin of, 193–195
seed dispersal, 27, 33–36, 37–38, 39
Sequoia National Forest and Giant Sequoia National Monument, 127–128
Serengeti National Park, 215–217
Sherwood Forest, 128
Sierra National Forest, 129–130
silviculture, 149–159
  artificial regeneration, 153–159
  natural regeneration, 152
soil conditions in forests, 3, 5

T

temperate forests, 4, 8, 53–75
  biological productivity of, 74–75
  biota of, 60–68
  development and structure of, 69–73
  environments of, 56–68
  human disturbance of, 73
  origin of, 54–56
temperate rainforests, 5, 7, 65
Teutoburg Forest, 130
thorn forests, 68

Thuringian Forest, 130–131
Tongass National Forest, 131–132
tree cover, influence of density of, 3
tree hollows, communities of animals in, 35, 67
tropical rainforests, 2, 4–7, 7–53, 145, 168
  about, 7–8
  biological productivity of, 41–42
  biota of, 13, 18–25, 33–40
  concerns for the future, 53
  conservation of, 52
  destruction/deforestation of, 13–14, 44–51
  and ecotourism, 52
  effects of population growth on, 45–46
  environments of, 14–18
  and evolution of humans, 13
  general structure of, 25–33
  origin of, 8–14
  and ranching and mining, 50–51
  relationships between flora and fauna, 33–40
  resettlement programs in, 46–49, 50
  soils of, 3, 5–6, 17–18
  status of the world's, 44–52
tundras, 178, 206–210
  alpine, 207, 210
  Arctic, 206, 207–210

U

urban forestry, 177

V

veld, 217–223
  animal life of, 222–223
  climate of, 220–221
  physiography of, 218–220
  plant life of, 221–222